Pitman Research Notes in Mathematics Series

Submission of proposals for consideration
Suggestions for publication, in the form of outlines and representative samples, are invited by
the Editorial Board for assessment. Intending authors should approach one of the main editors or
another member of the Editorial Board, citing the relevant AMS subject classifications.
Alternatively, outlines may be sent directly to the publisher's offices. Refereeing is by members
of the board and other mathematical authorities in the topic concerned, throughout the world.

Preparation of accepted manuscripts
On acceptance of a proposal, the publisher will supply full instructions for the preparation of
manuscripts in a form suitable for direct photo-lithographic reproduction. Specially printed grid
sheets can be provided and a contribution is offered by the publisher towards the cost of typing.
Word processor output, subject to the publisher's approval, is also acceptable.

Illustrations should be prepared by the authors, ready for direct reproduction without further
improvement. The use of hand-drawn symbols should be avoided wherever possible, in order to
maintain maximum clarity of the text.

The publisher will be pleased to give any guidance necessary during the preparation of a
typescript, and will be happy to answer any queries.

Important note
In order to avoid later retyping, intending authors are strongly urged not to begin final
preparation of a typescript before receiving the publisher's guidelines. In this way it is hoped to
preserve the uniform appearance of the series.

Longman Group Ltd
Longman House
Burnt Mill
Harlow, Essex, CM20 2JE
UK
(Telephone (0) 1279 426721)

Titles in this series. A full list is available from the publisher on request.

Shmuel Kantorovitz

Bar-Ilan University, Israel

Semigroups of operators and spectral theory

 **LONGMAN**

Copublished in the United States with
John Wiley & Sons Inc., New York.

Longman Group Limited
Longman House, Burnt Mill, Harlow
Essex CM20 2JE, England
and Associated companies throughout the world.

Copublished in the United States with
John Wiley & Sons Inc., 605 Third Avenue, New York, NY 10158

First published 1995

AMS Subject Classifications: (Main) 47D05, 47B40, 47A60
                             (Subsidiary) 47D10, 47A55, 47D40

ISSN 0269-3674

ISBN 0 582 27778 7

British Library Cataloguing in Publication Data

A catalogue record for this book is
available from the British Library

Library of Congress Cataloging-in-Publication Data

A catalog record for this book is available

Printed and bound in Great Britain
by Biddles Ltd, Guildford and King's Lynn

To Ita, Bracha, Pnina,

Pinchas, and Ruth.

TABLE of CONTENT

Introduction

PART I. GENERAL THEORY

PART II. GENERALIZATIONS

INTRODUCTION

These "Lecture Notes" were written for a second year graduate course on "Topics in Spectral Theory". They present some aspects of the theory of semigroups of operators, mostly from the point of view of its application to spectral theory, and even more specifically, to the integral representation of operators or families of operators. There is no attempt therefore to cover either the subject of "semigroups" or the subject of "spectral theory" thoroughly. These theories and their many applications to Differential Equations, Stochastic Processes, Mathematical Physics, etc..., have been the subject of many excellent books, such as [D], [Fat], [G],[HP],[Kat1], [P], [RS], and others. We refer the interested reader to these (and other) texts. Nevertheless, in order to permit a smooth reading of this monograph-type notes, and/or to make them convenient for a course or seminar, we have made them self contained by including a concise description of the basic facts on semigroups. The Hille-Yosida theory, concentrating on the concept of the *generator* (or *infinitesimal generator*) of the semigroup, is presented in Section A (Part I), culminating with the wellknown Hille-Yosida Theorem on the characterization of generators. A semigroup of operators is a function

$$T(.) : [0, \infty) \to B(X),$$

(where $B(X)$ denotes the Banach algebra of all bounded linear operators on a given Banach space X), such that $T(0)$ is the identity I and

$$T(s)T(t) = T(s + t) \qquad s, t \geq 0.$$

It is of class C_o if it is strongly right-continuous at 0. This implies strong continuity on $[0, \infty)$ and exponential growth. The generator A of $T(.)$ is essentially the right derivative at 0 with maximal domain $D(A)$. It is a closed densely defined operator, and for each $x \in D(A)$, the function $u = T(.)x$ is the unique solution of the Cauchy problem on $[0, \infty)$:

$$u' = Au; \qquad u(0) = x.$$

The generator is bounded if and only if $T(.)$ is uniformly continuous, and is then an ordinary exponential $T(t) = e^{tA}$. In general, the generator can be approximated pointwise on $D(A)$ by bounded operators

$$A_\lambda := \lambda[\lambda R(\lambda) - I]$$

(the so-called Hille-Yosida approximations of A), where $R(\lambda) := (\lambda I - A)^{-1}$ is the *resolvent* of A.

The Hille-Yosida theorem establishes that a closed densely defined operator A generates a C_o-semigroup with exponential growth $||T(t)|| \leq Me^{at}$ if and only if the resolvent exists for $\lambda > a$ and satisfies for all $m = 1, 2, ...$

$$||R(\lambda)^m|| \leq M(\lambda - a)^{-m}.$$

In the context of the Cauchy problem for A, the Hille-Yosida theorem characterizes the Cauchy problems that have a unique solution with exponential growth.

A characterization of generators that avoids resolvents uses the concept of dissipativity, introduced by Lumer and Phillips. This approach provides also an elegant perturbation theorem for generators, due to T. Kato. These matters are presented in Section C of Part I. In Section D, we prove the Trotter-Kato theorem about the equivalence of "generator graph convergence", "Strong resolvent convergence", and "Semigroup strong uniform convergence on compacta". Section E deals in the unified way due to T. Kato with the "exponential formula"

$$T(t) = \lim_n [\frac{n}{t} R(\frac{n}{t})]^n = \lim_n (I - \frac{tA}{n})^{-n}$$

(in the strong operator topology), and with the "Trotter Product Formula"

$$U(t) = \lim_n [S(t/n)T(t/n)]^n$$

(strongly), when $S(.), T(.), U(.)$ are C_o-semigroups generated respectively by A, B, and $A + B$.

The important Hille-Phillips perturbation theorem (that supplies a condition on a closed operator B that is sufficient for the perturbation $A + B$ of the generator A, to be also a generator) is proved in full detail in Section F.

The background material is concluded with a proof of the classical Stone theorem on unitary semigroups in Hilbert space (p. 38). As a prototype of integral representation theorems, Stone's theorem motivates our theory of the so-called "Semi-simplicity manifold" Z for a given group of operators (Section G). The linear manifold Z is defined by means of an adequate renorming, and it turns out to be maximal for the existence of a spectral integral representation of $T(.)$ on it (Theorem 1.49).

The renorming idea is also effective in creating the so-called Hille-Yosida Space (Section B), which is maximal with the property that the part of A in it generates a C_o-semigroup (cf. [K5]).

Section H touches upon the analyticity problem for semigroups, and gives a variant of a recent result of Liu with a (new) proof based on the exponential formula and normal families (cf [K8]).

Part I is concluded with the presentation of our recent "non-commutative Taylor formula" for semigroups as functions of their generator [K7].

Part II (pp. 65-114) describes some recent generalizations (published mostly after 1988) of the theory presented in Part I. Pre-semigroups (Section A), also called C-semigroups or regularized semigroups in the literature, have been introduced in germinal form in [DP], and their extensive study was started in [DPg]. Their main importance is in the solution of the Cauchy problem when A is not necessarily a generator (cf Theorem 2.5). The recent monograph [DL4] presents in detail many applications of this theory (and its extensions).

In Sections B and C, we extend the concept of the semi-simplicity manifold to apply to operators that are not necessarily generators, provided they have *real* spectrum (Section B), or at least have a half-line in their resolvent set (Section C). The construction is based as before on the renorming method. The operational calculus on the semi-simplicity manifold is developed for reflexive Banach space. A recent extension to the non-reflexive case is contained in [DLK1].

The related concepts of the Laplace-Stieltjes space and the Integrated Laplace space (cf. [DLK]) for a *family* of closed operators is defined in Section D, also by the renorming method, with application to the spectral integral representation of semigroups of *closed* operators, and to the characterization of generators of n-times integrated semigroups (cf [Neu] for the concept and its application to the Cauchy problem).

In Section E, we develop the Klein-Laundau theory of semigroups of *unbounded* symmetric operators, generalizing the classical Stone theorem (cf. [KL]). An analogous theory for cosine families of (unbounded) symmetric operators (cf. [KH3]) is presented in Section F. These theories provide a natural approach to Nelson's Analytic Vectors theorem and to Nussbaum's Semianalytic Vectors theorem, respectively. The Klein-Laundau theory has seen many applications to Mathematical Physics, but this subject is beyond the scope of these lectures.

PART I. GENERAL THEORY

A. THE HILLE-YOSIDA THEORY

THE GENERATOR.

Let X be a Banach space, and let $B(X)$ denote the Banach algebra of all bounded (linear) operators on X into X.

A function $T(.) : [0, \infty) \to B(X)$ is a semigroup if

$$T(s)T(t) = T(s + t) \qquad s, t \geq 0$$

and

$$T(0) = I,$$

where I denotes the identity operator.

The generator A of the semigroup $T(.)$ is the operator

$$Ax = \lim_{t \to 0+} [T(t)x - x]/t$$

with "maximal domain"

$$D(A) = \{x \in X; \quad above \quad limit \quad exists\}.$$

The above limit is the limit in X (with respect to the norm), and is in fact the strong derivative of $T(.)x$ at 0.

The "continuity at 0" (or C_o) condition is

$$\lim_{t \to 0+} T(t)x = x \quad for \quad all \quad x \in X.$$

This is continuity at zero in the strong operator topology on $B(X)$ (in brief, strong continuity at 0). This will be a fixed hypothesis.

1.1. THEOREM. Let $T(.)$ be a C_o-semigroup. Then it is strongly continuous on $[0, \infty)$, and there exist constants $M \geq 1$ and $a \geq 0$ such that

$$||T(t)|| \leq Me^{at}$$

for all $t \geq 0$.

PROOF. Let $c_n = \sup\{||T(t)||; t \in [0, 1/n]\}$ for $n = 1, 2, ...$

If $c_n = \infty$ for all n, there exist $t_n \in [0, 1/n]$ such that $||T(t_n)|| > n$ (for $n = 1, 2, 3, ...$). Then

$$\sup_n ||T(t_n)|| = \infty,$$

and so, by the Uniform Boundedness Theorem, there exists x such that

$$\sup_n ||T(t_n)x|| = \infty.$$

However the sequence $||T(t_n)x||$ converges to $||x||$ (by the C_o condition, since $t_n \to 0+$), and is therefore bounded. This contradiction shows that there exists an n for which $c_n < \infty$. Fix such an n, and let $c = c_n$. Note that $c \geq ||T(0)|| = ||I|| = 1$.

For any $t > 0$, the semigroup property gives

$$T(t) = T(1/n)^{n[t]} T(\{t\}/n)^n,$$

where $[t]$ denotes the entire part of t, and $\{t\}$ its fractional part. Since $1/n$ and $\{t\}/n$ are both in $[0, 1/n]$, we have $||T(1/n)|| \leq c$ and $||T(\{t\}/n)|| \leq c$, so that $||T(t)|| \leq (c^n)^{[t]+1} \leq (c^n)^{t+1} = Me^{at}$, where $M = c^n \geq 1$ and $a = n \log c \geq 0$ (we used the fact that $c \geq 1$).

Continuity at $t > 0$. (1) For $h > 0$, we have for all $x \in X$

$$||T(t+h)x - T(t)x|| = ||T(h)[T(t)x] - [T(t)x]|| \to 0$$

as $h \to 0$, by the C_o condition with the fixed vector $T(t)x$.

(2) For $h < 0$, write $h = -k$, with $0 < k < t$. Then

$$||T(t+h)x - T(t)x|| = ||T(t-k)(x - T(k)x)||$$

$$\leq Me^{a(t-k)}||T(k)x - x|| \to 0$$

as $h \to 0$ by the C_o property.||||

1.2. THEOREM. Let A be the generator of the C_o-semigroup $T(.)$. Then:
1. A is closed and densely defined.
2. For each $t \geq 0$, $T(t)D(A) \subset D(A)$, and

$$AT(t)x = T(t)Ax = (d/dt)[T(t)x]$$

for each $x \in D(A)$.

3. For each $x \in D(A)$, the function $u = T(.)x$ is C^1 on $[0, \infty)$, and is the unique solution of the "Abstract Cauchy Problem" (ACP) on $[0, \infty)$:

$$du/dt = Au; \qquad u(0) = x.$$

4

PROOF. For each given $x \in X$, the function $T(.)x$ is continuous on $[0, \infty)$, by Theorem 1.1, and has therefore a Riemann integral over any finite interval $[0, t]$. Denote this integral by x_t. Also let $A_h = [T(h) - I]/h$ for $h > 0$. Then

$$A_h x_t = h^{-1}[\int_0^t T(s + h)x\,ds - \int_0^t T(s)x\,ds]$$

$$= h^{-1}[(\int_h^{t+h} - \int_0^t)T(s)x\,ds$$

$$= h^{-1}\int_t^{t+h} T(s)x\,ds - h^{-1}\int_0^h T(s)x\,ds$$

$$\to T(t)x - x$$

as $h \to 0+$, by continuity of $T(.)x$.

Hence $x_t \in D(A)$ and
$$Ax_t = T(t)x - x. \tag{$*$}$$

The C_o-condition implies that $x_t/t(\in D(A)) \to x$, and therefore $D(A)$ is dense in X.

If $x \in D(A)$, then for each $t > 0$,
$$A_h T(t)x = T(t)A_h x \to T(t)Ax \tag{$**$}$$

as $h \to 0+$. Hence $T(t)x \in D(A)$ and

$$AT(t)x = T(t)Ax.$$

The left hand side in $(**)$ is also equal to $h^{-1}[T(t + h)x - T(t)x]$, and so the right derivative of $T(.)x$ exists, is equal to $A[T(.)x] = T(.)(Ax)$, and is in particular continuous.

If $0 < k < t$, we have for $x \in D(A)$

$$||(-k)^{-1}[T(t - k)x - T(t)x] - T(t)Ax||$$

$$\leq ||T(t - k)(A_k x - Ax)|| + ||T(t - k)Ax - T(t)Ax||$$

$$\leq Me^{a(t-k)}||A_k x - Ax|| + ||T(t - k)Ax - T(t)Ax|| \to 0$$

as $k \to 0+$, since $x \in D(A)$ and $T(.)$ is strongly continuous.

Thus $u = T(.)x$ is of class C^1 on $[0, \infty)$, and solves ACP.

Suppose $v : [0, \infty) \to D(A)$ is differentiable. Then

$$h^{-1}[T(t+h)v(t+h) - T(t)v(t)] =$$

$$T(t)A_h v(t) + T(t+h)[h^{-1}(v(t+h) - v(t)) - v'(t)] + T(t+h)v'(t).$$

The first term on the right has limit $T(t)Av(t)$ when $h \to 0$, since $v(t) \in D(A)$ and $T(t) \in B(X)$. The second term has limit 0, since $||T(t+h)|| \leq Me^{a(t+h)}$. The last term has limit $T(t)v'(t)$, by strong continuity of $T(.)$. Hence $(d/dt)[T(t)v(t)] = T(t)[Av(t) + v'(t)]$. Suppose now that v solves ACP with a given $x \in D(A)$, in some interval $[0, \tau]$. Fix $s \in (0, \tau]$. Then (by the fundamental theorem of calculus):

$$T(s)x - v(s) = \int_0^s (d/dt)[T(t)v(s-t)]dt$$

$$= \int_0^s T(t)[Av(s-t) - v'(s-t)]dt = 0,$$

which proves the uniqueness.

By the fundamental theorem of calculus and (*),

$$A \int_0^t T(s)x\,ds = T(t)x - x = \int_0^t T(s)Ax\,ds$$

for $t > 0$ and $x \in D(A)$. Suppose $x_n \in D(A)$ are such that $x_n \to x$ and $Ax_n \to y$ in X. If $V(.) : [0, \tau] \to B(X)$ is strongly continuous, then $||V(.)||$ is a bounded measurable function and for each $x \in X$,

$$\left|\left| \int_0^\tau V(t)x\,dt \right|\right| \leq \int_0^\tau ||V(t)||dt||x||$$

(see below).

Therefore, as $n \to \infty$,

$$\left|\left| \int_0^t T(s)Ax_n\,ds - \int_0^t T(s)y\,ds \right|\right| \leq \int_0^t ||T(s)||ds||Ax_n - y|| \leq const.||Ax_n - y|| \to 0.$$

Hence

$$A_t x = \lim_n A_t x_n = \lim_n t^{-1} \int_0^t T(s)Ax_n = t^{-1} \int_0^t T(s)y\,ds \to y$$

as $t \to 0+$.

This shows that $x \in D(A)$ and $Ax = y$, i.e., A is closed.

Back to the claim about $V(.)$, the boundedness of $||V(.)||$ follows immediately from the strong continuity of $V(.)$ and the Uniform Boundedness Theorem. To

6

prove the measurability of $||V(.)||$, it suffices to show that the set $C = \{t \in [0, \tau]; ||V(t)|| > c\}$ is Borel for each $c \geq 0$. If $t \in C$, there exists $x \in X$ with norm 1 such that $||V(t)x|| > c$, and by continuity of $||V(.)x||$, there is a neighborhood of t in $[0, \tau]$ where $||V(.)x|| > c$, and so $||V(.)|| > c$ there. Hence C is open, so certainly Borel. We got actually that $||V(.)||$ is lower semicontinuous (which is stronger than Borel measurability).||||

TYPE AND SPECTRUM.

By Theorem 1.1, $\log ||T(.)||$ is bounded above on finite intervals and clearly subadditive. We need the following general lemma on such functions.

1.3. LEMMA. Let $p : [0, \infty) \to [-\infty, \infty)$ be subadditive (i.e., $p(t+s) \leq p(t)+p(s)$ for all t, s in the domain of p) and bounded above in $[0, 1]$. Then

$$-\infty \leq \inf_{t>0} \frac{p(t)}{t} = \lim_{t\to\infty} \frac{p(t)}{t} < \infty.$$

PROOF. If $p(t_0) = -\infty$ for some t_0, then for all $t \geq t_0$, $p(t) \leq p(t_0) + p(t - t_0) = -\infty$, and the result is trivial. So we may assume that p is finite. Fix $s > 0$ and $r > p(s)/s$. For $t > 0$ arbitrary, let n be the unique positive integer such that $ns \leq t < (n+1)s$. Then

$$p(t)/t = p(ns + (t - ns))/t \leq np(s)/t + p(t - ns)/t$$

$$< rns/t + \sup_{[0,s]} p/t.$$

Since the hypothesis imply that p is bounded above on any interval $[0, s]$, it follows that $\limsup_{t\to\infty} \frac{p(t)}{t} \leq r$, for any $r > p(s)/s$. Hence

$$\limsup \frac{p(t)}{t} \leq \inf_{s>0} \frac{p(s)}{s} \leq \liminf \frac{p(t)}{t},$$

and the lemma follows.||||

In particular, the *type* of $T(.)$ is (fixed notation!)

$$\omega := \inf_{t>0} \frac{\log ||T(t)||}{t} = \lim_{t\to\infty} \frac{\log ||T(t)||}{t}.$$

For any non-negative $a > \omega$, we clearly have

$$||T(t)|| \leq Me^{at}$$

for all $t \geq 0$ (where the constant $M \geq 1$ depends on a).

1.4. THEOREM. The spectral radius of $T(t)$ is $e^{\omega t}$.

PROOF. Since the claim is trivial for $t = 0$, fix $t > 0$, and let $r(T(t))$ denote the spectral radius of $T(t)$. By the Beurling-Gelfand formula and Lemma 1.3, we have

$$r(T(t)) = \lim_{n} ||T(t)^n||^{1/n} = \lim_{n} e^{(1/n)\log ||T(nt)||}$$

$$= e^{t \lim_n (1/nt) \log ||T(nt)||} = e^{\omega t}.||||$$

UNIFORM CONTINUITY.

The next theorem shows that the stronger hypothesis of continuity at zero in the uniform operator topology (that is, in the norm topology of $B(X)$) yields to a rather uninteresting class of semigroups.

1.5. THEOREM. The semigroup $T(.)$ is norm-continuous at 0 iff its generator A belongs to $B(X)$; in that case, $T(t) = e^{tA}$ (defined as the usual power series, which converges in $B(X)$).

PROOF. 1. If $A \in B(X)$, one verifies directly that e^{tA} is a well-defined norm-continuous group with generator A. Since by Theorem 1.2 the generator determines the semigroup uniquely, and A is also the generator of $T(.)$, we have $T(t) = e^{tA}$, so that, in particular, $T(.)$ is norm-continuous.

2. Suppose conversely that $T(.)$ is norm-continuous at 0 (hence everywhere on $[0, \infty)$, by the argument in the proof of Theorem 1.1). We may then consider Riemann integrals of $T(.)$, defined as the usual limits (in $B(X)$!). For $h, t > 0$, a calculation as at the beginning of the proof of Theorem 1.2 shows that

$$[T(t) - I] \int_0^h T(s)ds = [T(h) - I] \int_0^t T(s)ds.$$

8

Since $||h^{-1} \int_0^h T(s)ds - I|| \to 0$ when $h \to 0+$ by norm-continuity of $T(.)$, we can fix h so small that the above norm is less than 1, and therefore $V := \int_0^h T(s)ds$ is invertible in $B(X)$. Hence,

$$T(t) - I = \int_0^t T(s)ds.A,$$

where $A := [T(h) - I]V^{-1} \in B(X)$ (the change of order in the calculation is valid, since the values of $T(.)$ commute). Dividing by t and letting $t \to 0$, we get $t^{-1}[T(t) - I] \to A$ in $B(X)$, by norm continuity of $T(.)$. Hence $A(\in B(X)!)$ is the generator of $T(.)$.||||

CORE FOR THE GENERATOR

1.6. Let A be any closed operator with domain $D(A)$ in X. The "graph-norm" on $D(A)$ is the norm
$$|x|_A := ||x|| + ||Ax||$$
induced on the graph of A by the norm on X^2. $D(A)$ is a Banach space under the graph-norm (because A is closed), and we shall use the notation $[D(A)]$ for this Banach space. Any subspace D_0 dense in $[D(A)]$ is called a "core" for A. Explicitely, a subspace D_0 of D(A) is a core for A iff for any $x \in D(A)$, there exists a sequence $\{x_n\}$ in D_0 such that $x_n \to x$ and $Ax_n \to Ax$ (i.e., A equals the closure $(A/D_0)^-$ of its restriction to D_0).

Since it is often difficult to determine $D(A)$, it is important (and sufficient in most case) to know a core for A. The following theorem gives a simple useful tool in this direction for the generator A of the semigroup $T(.)$.

1.7. THEOREM. If D_0 is a subspace of $D(A)$ dense in X and $T(.)$- invariant, then it is a core for A.

PROOF. Note first that $T(.)$ is a C_o-semigroup in the Banach space $[D(A)]$, since for all $x \in D(A)$, when $t \to 0+$,

$$|T(t)x - x|_A = ||T(t)x - x|| + ||T(t)(Ax) - (Ax)|| \to 0.$$

Therefore, for $x \in D_0$, Riemann integrals (over finite intervals) of $T(.)x$ make sense in the graph-norm, and belong to D_0^-, the closure of D_0 in $[D(A)]$. Let $x \in D(A)$. By density of D_0 in X, there exists a sequence $\{x_n\} \subset D_0$ such that $x_n \to x$ in X.

9

The elements x_t and $(x_n)_t$ (see notation in proof of Theorem 1.2) are in $D(A)$, and for each $t > 0$

$$|(x_n)_t - x_t|_A = ||\int_0^t T(s)(x_n - x)ds||$$

$$+ ||[T(t)x_n - x_n] - [T(t)x - x]|| \to 0$$

when $n \to \infty$. Since $(x_n)_t \in D_0^-$ for each n, we have also $x_t \in D_0^-$. Finally, by the C_0-property of $T(.)$ in $[D(A)]$, $t^{-1}x_t(\in D_0^-!) \to x$ in the graph-norm, and so $x \in D_0^-$.||||

A useful core for A is the space $D^\infty = D^\infty(A)$ of all "C^∞-vectors" for A, that is, the set of all $x \in X$ for which the function $T(.)x$ is of class C^∞ on $[0, \infty)$.

1.8. THEOREM. 1. $D^\infty = \bigcap_{n=1}^\infty D(A^n)$.
2. D^∞ is dense in X and $T(.)$-invariant.
3. D^∞ is a core for A.

PROOF. 1. and 2. imply 3. by Theorem 1.7.

If $x \in D^\infty$, $T(.)x$ is differentiable at 0, i.e., $x \in D(A)$, and $(d/dt)T(t)x = T(t)(Ax)$. Hence $Ax \in D^\infty$, and so, in particular, $x \in D(A^2)$. Inductively, $x \in D(A^n)$ and

$$(*) \qquad [T(.)x]^{(n)} = T(.)A^n x$$

for all $n = 1, 2, 3....$ Conversely, if $x \in D(A^n)$ for all n, then $T(.)x$ is differentiable and $[T(.)x]' = T(.)Ax$ (cf. Theorem 1.2), so that, inductively, we obtain that $T(.)x$ is of class C^∞ and $(*)$ is valid. This proves 1. and the $T(.)$- invariance of D^∞. To prove the density of D^∞, we use an "approximate identity" $0 \le h_n \in C^\infty$ with support in $(0, 1/n)$ and integral (over $\mathbb{R}$) equal to 1. Given $x \in X$, define $x_n = \int_0^\infty h_n(t)T(t)x dt$. Then $x_n \to x$ in X. It remains to show that $x_n \in D^\infty$ for all n. For $k > 0$,

$$A_k x_n = k^{-1} \int_0^\infty h_n(t)[T(t+k)x - T(t)x]dt = \int_0^\infty k^{-1}[h_n(t-k) - h_n(t)]T(t)x dt$$

$$\to - \int_0^\infty h_n'(t)T(t)x dt$$

when $k \to 0+$. Hence $x_n \in D(A)$ and $Ax_n = - \int_0^\infty h_n'(t)T(t)x dt$. Repeating the argument, we obtain $x_n \in D(A^j)$ for all j and $A^j x_n = (-1)^j \int_0^\infty h_n^{(j)}(t)T(t)x dt$. The conclusion follows now from 1.||||

THE RESOLVENT

The verification of the following elementary facts is left as an exercise.

1.9. PROPOSITION. Let A be a closed operator, with domain $D(A)$. Then: 1. If A is bijective, its inverse with domain $D(A^{-1})$ equal to the range $ran\,(A)$ of A, is closed.
2. If $B \in B(X)$ and $\alpha, \beta \in \mathbb{C}$, then $\alpha A + \beta B$, with domain X for $\alpha = 0$ and $D(A)$ otherwise, is closed.
3. If $B \in B(X)$, then AB, with its maximal domain, is closed. If B is non-singular, then BA, with domain $D(A)$, is also closed.

1.10. DEFINITION. The "resolvent set" $\rho(A)$ of the closed operator A is the set of all complex λ for which $\lambda I - A$ is bijective (i.e., one-to-one and onto X). Its complement is the "spectrum" $\sigma(A)$ of A.

The operator $R(\lambda) = R(\lambda; A) := (\lambda I - A)^{-1}$ for $\lambda \in \rho(A)$ is closed (see 1.9) and everywhere defined, and belongs therefore to $B(X)$ by the closed graph theorem. It is called the "resolvent of A". It is useful to observe that $\lambda \in \rho(A)$ iff there exists an operator $R(\lambda) \in B(X)$ with range in $D(A)$ such that

$$(\lambda I - A)R(\lambda)x = x \qquad (x \in X)$$

and

$$R(\lambda)(\lambda I - A)x = x \qquad (x \in D(A)).$$

It is useful to write the above relations in the form

$$R(\lambda)A \subset AR(\lambda) = \lambda R(\lambda) - I \qquad (*)$$

(where all operators are with their maximal domain).

1.11. THEOREM. Let A be a closed operator. Then $\rho(A)$ is open, $R(.)$ is analytic on $\rho(A)$ and satisfies the "resolvent equation"

$$R(\lambda) - R(\mu) = (\mu - \lambda)R(\lambda)R(\mu).$$

Also $||R(\lambda)|| \geq \frac{1}{d(\lambda, \sigma(A))}$.

PROOF. Let $\lambda \in \rho(A)$, and set $\delta = ||R(\lambda)||^{-1}$. The series

$$S(\zeta) = \Sigma_{n \geq 0}(-1)^n R(\lambda)^{n+1}(\zeta - \lambda)^n$$

is norm-convergent in $B(X)$ for $|\zeta - \lambda| < \delta$, and so defines an element of $B(X)$. For $x \in D(A)$,

$$S(\zeta)(\zeta I - A)x = S(\zeta)[(\zeta - \lambda)I + (\lambda I - A)]x$$

$$= \Sigma(-1)^n R(\lambda)^{n+1}(\zeta - \lambda)^{n+1}x$$

$$+ \Sigma(-1)^n R(\lambda)^n(\zeta - \lambda)^n x = x.$$

Next, for any $x \in X$, let x_m denote the m-th partial sum of the series $S(\zeta)x$. Then $x_m \in D(A)$ (because $x_m \in ran\, R(\lambda) = D(A)$), $x_m \to S(\zeta)x$, and by $(*)$

$$Ax_m = \Sigma_{0 \le n \le m}(-1)^n \lambda R(\lambda)^{n+1}(\zeta - \lambda)^n x$$

$$- \Sigma_{0 \le n \le m}(-1)^n R(\lambda)^n(\zeta - \lambda)^n x$$

$$\to \lambda S(\zeta)x + (\zeta - \lambda)S(\zeta)x - x$$

$$= \zeta S(\zeta)x - x.$$

Since A is closed, it follows that $S(\zeta)x \in D(A)$ and $(\zeta I - A)S(\zeta)x = x$, and we conclude that $\zeta \in \rho(A)$ and $R(\zeta) = S(\zeta)$ for all ζ such that $|\zeta - \lambda| < \delta$. Hence $\rho(A)$ is open and $R(.)$ is analytic on $\rho(A)$. Also, since the disc of radius δ around λ is contained in $\rho(A)$, we have $d(\lambda, \sigma(A)) \ge \delta := \|R(\lambda)\|^{-1}$.

Finally, for $\lambda, \mu \in \rho(A)$ and $x \in X$,

$$(\lambda I - A)[R(\lambda) - R(\mu) - (\mu - \lambda)R(\lambda)R(\mu)]x$$

$$= x - [(\lambda - \mu)I + (\mu I - A)]R(\mu)x - (\mu - \lambda)R(\mu)x$$

$$= x - (\lambda - \mu)R(\mu)x - x + (\lambda - \mu)R(\mu)x = 0.$$

Since $\lambda I - A$ is injective, the resolvent equation follows. ‖‖‖

It will be convenient to consider the following general concept.

1.12. DEFINITION. A "pseudo-resolvent" is a function $R(.)$, defined on an open set $U \subset \mathbb{C}$, with values in $B(X)$, and satisfying the resolvent equation in U.

1.13. THEOREM. If $R(.) : U \to B(X)$ is a pseudo-resolvent, then $ker\, R(\lambda)$ and $ran\, R(\lambda)$ are independent of $\lambda \in U$, and $R(.)$ is the resolvent of some closed A with $U \subset \rho(A)$ iff $ker\, R(\lambda) = 0$.

PROOF. Let $\lambda, \mu \in \rho(A)$. If $x \in ker\, R(\lambda)$, we have by the resolvent equation $R(\mu)x = R(\lambda)x + (\lambda - \mu)R(\mu)R(\lambda)x = 0$, i.e., $x \in ker\, R(\mu)$, and so, by symmetry, $ker\, R(\lambda) = ker\, R(\mu)$.

12

If $y \in ran\, R(\lambda)$, write $y = R(\lambda)x$, and then $y = R(\mu)[x + (\mu - \lambda)R(\lambda)x] \in ran\, R(\mu)$, so that $ran\, R(\lambda) = ran\, R(\mu)$ by symmetry.

Suppose $ker\, R(\lambda) = 0$ for some (hence for all)$\lambda \in U$. Then $A := \lambda I - R(\lambda)^{-1} : ran\, R(\lambda) \to X$ is closed, and since $\lambda I - A = R(\lambda)^{-1}$ and $R(\lambda) \in B(X)$, the operator $\lambda I - A$ is bijective. Thus $\lambda \in \rho(A)$ and $R(\lambda) = (\lambda I - A)^{-1}$.

For any $\mu \in U$, we have by the resolvent equation

$$R(\mu) = R(\lambda)[I + (\lambda - \mu)R(\mu)]$$

and therefore

$$(\mu I - A)R(\mu) = (\mu - \lambda)R(\mu) + (\lambda I - A)R(\lambda)[I + (\lambda - \mu)R(\mu)] = I$$

and similarly $R(\mu)(\mu I - A) \subset I$. Therefore $\mu \in \rho(A)$ and $R(\mu; A) = R(\mu)$.

Conversely, if $R(\lambda) = R(\lambda; A)$ for all $\lambda \in U$ (for some closed operator A), then $R(.)$ is a pseudo-resolvent by Theorem 1.10, $U \subset \rho(A)$, and $ker\, R(\lambda) = 0$ trivially.||||

Another characterization of resolvents among pseudo-resolvents uses the range of $R(\lambda)$.

1.14. THEOREM. Let $R(.) : (\omega, \infty) \to B(X)$ be a pseudo-resolvent such that for all $\lambda > \omega$,

$$||R(\lambda)|| \leq \frac{M}{\lambda - \omega}. \tag{*}$$

Then $R(.)$ is the resolvent of a closed densely-defined operator iff the range of $R(\lambda)$ is dense in X for some (hence for all)λ.

PROOF. The necessity is trivial, since $ran\, R(\lambda; A) = D(A)$.

Sufficiency. For $x \in ran\, R(\lambda)$, write $x = R(\lambda)y$, and then, for $\mu > \omega$,

$$\mu R(\mu)x = \mu R(\mu)R(\lambda)y = \frac{\mu}{\mu - \lambda}[R(\lambda)y - R(\mu)y] \to R(\lambda)y = x$$

when $\mu \to \infty$, by the growth condition. Since $ran\, R(\lambda)$ is dense in X, it follows from the growth condition that $\mu R(\mu)x \to x$ for all $x \in X$

[indeed, let $x_n \in ran\, R(\lambda)$ converge to x. Then

$$||\mu R(\mu)x - x|| \leq ||\mu R(\mu)x_n - x_n|| + ||\mu R(\mu) - I||.||x - x_n||.$$

The second term on the right is $\leq [M\mu/(\mu - \omega) + 1]||x - x_n|| \to (M + 1)||x - x_n||$, and the first term $\to 0$ when $\mu \to \infty$ (for each fixed n).Therefore

$$\limsup_{\mu \to \infty} ||\mu R(\mu)x - x|| \leq (M + 1)||x - x_n||.$$

13

Letting $n \to \infty$, the conclusion follows].

Suppose $x \in \ker R(\lambda)$ for some $\lambda > \omega$. Then $x \in \ker R(\mu)$ for all $\mu > \omega$, but then $x = \lim \mu R(\mu)x = 0$, i.e., $\ker R(\lambda) = 0$, and so $R(\lambda) = R(\lambda; A)$ with A closed (by Theorem 1.13) and $D(A) = ran\, R(\lambda)$ dense, by hypothesis.||||

LAPLACE TRANSFORM.

We show next that the resolvent of A is the Laplace transform of $T(.)$.

1.15. THEOREM. 1. $\sigma(A) \subset \{\lambda \in \mathbb{C}; \Re\lambda \leq \omega\}$.
2. For $\Re\lambda > \omega$ and $x \in X$,

$$R(\lambda)x = \int_0^\infty e^{-\lambda t} T(t)x\,dt.$$

3. For $c > \omega, t > 0$ and $x \in D(A)$,

$$T(t)x = \lim_{\tau \to \infty} \frac{1}{2\pi i} \int_{c-i\tau}^{c+i\tau} e^{\lambda t} R(\lambda; A)x\,d\lambda,$$

where the limit is a strong limit in X.

PROOF. For any $a > \omega$, $\|T(t)\| = O(e^{at})$, and therefore the Laplace integral $L(\lambda)x$ defined in 2. converges absolutely for $\Re\lambda > a$, and defines an operator $L(\lambda) \in B(X)$ satisfying

$$\|L(\lambda)\| \leq \frac{M}{\Re\lambda - a}.$$

If $x \in D(A)$,

$$L(\lambda)(\lambda I - A)x = \int_0^\infty \{\lambda e^{-\lambda t} T(t)x - e^{-\lambda t}[T(t)x]'\}dt$$

$$= -\int_0^\infty [e^{-\lambda t} T(t)x]'\,dt = x.$$

On the other hand, for any $x \in X$ and $h > 0$,

$$A_h L(\lambda)x = h^{-1} \int_0^\infty e^{-\lambda t}[T(t+h)x - T(t)x]dt$$

$$= h^{-1}(e^{\lambda h} - 1)L(\lambda)x - e^{\lambda h}h^{-1}\int_0^h e^{-\lambda t}T(t)x ds$$

$$\rightarrow_{h \to 0+} \lambda L(\lambda)x - x.$$

Since A is closed, it follows that $L(\lambda)X \subset D(A)$ and $(\lambda I - A)L(\lambda)x = x$ for all $x \in X$, and we conclude that $L(\lambda) = R(\lambda; A)$ for all λ in the half-plane $\Re\lambda > a$. Since $a > \omega$ was arbitrary, Statements 1. and 2. are proved.

To obtain 3., we observe that $T(.)x$ is of class C^1 on $[0, \infty)$ (by Theorem 1.2), and we may therefore apply the (vector version of the) classical Complex Inversion Theorem for the Laplace transform (cf. Theorem 7.3 in [W]). ‖‖‖

The Laplace integral representation of $R(\lambda; A)$ implies the growth condition

$$||R(\lambda; A)|| \le \frac{M}{\lambda - a} \qquad (*)$$

for all $\lambda > a$ (where $a > \omega$ is fixed). Consider now *any* closed densely defined operator A with $(a, \infty) \subset \rho(A)$, which satisfies (*) for all $\lambda > \lambda_0$ (for some $\lambda_0 \ge a$). For short, call such an operator an **abstract potential**.

1.16. LEMMA. Let A be an abstract potential, and consider the bounded operators

$$A_\lambda := \lambda A R(\lambda) = \lambda[\lambda R(\lambda) - I]$$

for $\lambda > a$. Then as $\lambda \to \infty$,
  1. $A_\lambda x \to Ax$ for all $x \in D(A)$;
  2. $\lambda R(\lambda) \to I$ strongly (equivalently, $AR(\lambda) \to 0$ strongly).

PROOF. For $x \in D(A)$ and $\lambda > \lambda_0$,

$$||AR(\lambda)x|| = ||R(\lambda)Ax|| \le \frac{M}{\lambda - a}||Ax|| \to 0.$$

Since

$$||AR(\lambda)|| = ||\lambda R(\lambda) - I|| \le \frac{\lambda M}{\lambda - a} + 1 = O(1)$$

when $\lambda \to \infty$, and since $D(A)$ is dense in X, it follows that

$$AR(\lambda)x \to 0$$

for all $x \in X$. This is equivalent to 2.

Next, for $x \in D(A)$,
$$A_\lambda x = \lambda R(\lambda)(Ax) \to Ax$$

by 2.‖‖‖

Note that the notation A_λ in the present context should not be confused with the notation A_h used in previous sections.

When A is the generator of a semigroup $T(.)$ satisfying $||T(t)|| \le Me^{at}$, the growth property (*) can be strengthened as follows:
For any finite set of $\lambda_k > a$, $\quad k = 1, ..., m$,

$$||\Pi_k(\lambda_k - a)R(\lambda_k; A)|| \le M. \tag{**}$$

In particular (with all λ_k equal λ),

$$||R(\lambda; A)^m|| \le \frac{M}{(\lambda - a)^m} \tag{***}$$

for all $\lambda > a$ and $m = 1, 2, 3, ...$

Indeed, for all $x \in X$,
$$||\Pi_k(\lambda_k - a)R(\lambda_k; A)x||$$

$$= || \int_0^\infty ... \int_0^\infty \Pi_k(\lambda_k - a)e^{-\lambda_1 t_1 - ... - \lambda_m t_m} T(t_1 + ... + t_m)x \, dt_1 ... dt_m||$$

$$\le M \int_0^\infty ... \int_0^\infty \Pi_k(\lambda_k - a)e^{-(\lambda_k - a)t} dt ||x||$$

$$= M\Pi_k \int_0^\infty (\lambda_k - a)e^{-(\lambda_k - a)t} dt ||x|| = M||x||.$$

We prove now that (***) characterizes *generators* among all abstract potentials.

THE HILLE-YOSIDA THEOREM.

1.17. THEOREM. An operator A is the generator of a C_o-semigroup $T(.)$ (satisfying $||T(t)|| \le Me^{at}$ for all $t \ge 0$) iff

(1) it is closed and densely defined; and

(2) $(a, \infty) \subset \rho(A)$ and (***) is valid.

PROOF. We already saw the necessity of (1) and (2). Let then A satisfy (1) and (2). In particular, it is an abstract potential, and so Lemma 1.16 is satisfied. Define

$$T_\lambda(t) = e^{tA_\lambda}.$$

We have for $\lambda > 2a$ (so that $\frac{a\lambda}{\lambda - a} < 2a$):

$$||T_\lambda(t)|| \le e^{-\lambda t} \Sigma_n \frac{t^n \lambda^{2n}}{n!} ||R(\lambda)^n|| \le M e^{-\lambda t} \Sigma_n \frac{t^n \lambda^{2n}}{n!(\lambda - a)^n} = M e^{t \frac{a\lambda}{\lambda - a}} \le M e^{2at}.$$

Also for $\lambda \to \infty$,

$$\limsup ||T_\lambda(t)|| \le M e^{at}. \qquad (1)$$

CLAIM: $T_\lambda(t)$ converge in the strong operator topology (as $\lambda \to \infty$), uniformly for t in bounded intervals.

For $x \in D(A)$ and $\lambda, \mu > 2a$,

$$||T_\mu(t)x - T_\lambda(t)x|| = ||\int_0^t (d/ds)[T_\lambda(t - s)T_\mu(s)x]ds||$$

$$= ||\int_0^t T_\lambda(t - s)T_\mu(s)(A_\mu - A_\lambda)x ds|| \le M^2 e^{4at} t ||A_\mu x - A_\lambda x|| \to 0$$

when $\lambda, \mu \to \infty$, by Lemma 1.16, uniformly for t in bounded intervals.

Since $||T_\lambda(.)||$ is uniformly bounded in bounded intervals (by (1)), it follows from the density of $D(A)$ that $\{T_\lambda(t)x\}$ is Cauchy (as $\lambda \to \infty$) for all $x \in X$, uniformly for t in bounded intervals.

Define therefore

$$T(t)x = \lim_{\lambda \to \infty} T_\lambda(t)x$$

for $x \in X$ (limit in X-norm).

By (1), $||T(t)|| \le M e^{at}$ for all $t \ge 0$. The semigroup property of $T(.)$ follows from that of $T_\lambda(.)$. The uniform convergence on bounded intervals implies the continuity of $T(.)x$ on $[0, \infty)$, for each $x \in X$. Let A' denote the generator of $T(.)$. We have

$$T_\lambda x - x = \int_0^t T_\lambda(s)A_\lambda x ds.$$

For $x \in D(A)$, Lemma 1.16 implies (by letting $\lambda \to \infty$)

$$T(t)x - x = \int_0^t T(s)Ax ds.$$

Dividing by $t > 0$ and letting $t \to 0+$, we conclude that $x \in D(A')$ and $A'x = Ax$. Thus, for $\lambda > a$, $\lambda I - A$ and $\lambda I - A'$ are both one-to-one and onto X, and coincide on $D(\lambda I - A) = D(A)$. Therefore $D(A) = D(A')$, and the proof is complete. $||||$

For contraction semigroups (i.e., $||T(.)|| \le 1$), the Hille-Yosida characterization is especially simple (case $M = 1, a = 0$).

1.18. COROLLARY. An operator A is the generator of a C_o- contraction semigroup iff it is closed, densely defined, and $\lambda R(\lambda; A)$ (exist and) are contractions for all $\lambda > 0$.

We call the bounded operators A_λ the **Hille-Yosida approximations of** A. From Lemma 1.16 and the proof of the Hille-Yosida theorem, $A_\lambda x \to Ax$ for all $x \in D(A)$ and $e^{tA_\lambda} \to T(t)$ strongly, uniformly on bounded t-intervals (as $\lambda \to \infty$).

B. THE HILLE-YOSIDA SPACE

The inequalities (**) following the proof of Lemma 1.16 can be used to construct, for an arbitrary (unbounded) operator A with $(a, \infty) \subset \rho(A)$, a maximal Banach subspace Z of X such that A_Z, the "part of A in Z", generates a C_o-semigroup in Z.

1.19. DEFINITION. A Banach subspace Y of X is a linear manifold $Y \subset X$ which is a Banach space for a norm $||.||_Y \geq ||.||$.

If A is any operator on X with domain $D(A)$, and W is a linear manifold in X, the "part of A in W", denoted A_W, is the restriction of A to its maximal domain as an operator in W:

$$D(A_W) = \{x \in D(A); x, Ax \in W\}.$$

1.20. DEFINITION. Let A be an arbitrary operator with $(a, \infty) \subset \rho(A)$ for some real a. Denote

$$||x||_Y = \sup ||\Pi_k(\lambda_k - a)R(\lambda_k; A)x||,$$

where the supremum is taken over all finite subsets $\{\lambda_1, ..., \lambda_m\}$ of (a, ∞) (the product over the empty set is defined as x). Set

$$Y = \{x \in X; ||x||_Y < \infty\}.$$

1.21. LEMMA. The space $Y = (Y, ||.||_Y)$ is a Banach subspace of X, invariant under any bounded operator U which commutes with A, and $||U||_{B(Y)} \leq ||U||_{B(X)}$.

PROOF. Clearly, Y is a linear manifold in X, and its norm majorizes $||.||$. In particular, if $\{x_n\}$ is Cauchy in Y, it is also Cauchy in X; let x be its X-limit, and let $K = \sup_n ||x_n||_Y$. For any finite set $\{\lambda_k\}_{1 \leq k \leq m} \subset (a, \infty)$,

$$||\Pi_k(\lambda_k - a)R(\lambda_k; A)x|| = \lim_n ||\Pi_k(\lambda_k - a)R(\lambda_k; A)x_n|| \leq \limsup_n ||x_n||_Y \leq K,$$

so that $||x||_Y \leq K < \infty$, i.e., $x \in Y$.

Given $\epsilon > 0$, there exists n_o such that $||x_n - x_p||_Y < \epsilon$ whenever $n, p > n_o$. Therefore for any finite set $\{\lambda_k\}$ as before,

$$||\Pi_k(\lambda_k - a)R(\lambda_k; A)(x_n - x_p)|| \leq ||x_n - x_p||_Y < \epsilon$$

if $n, p > n_o$. Letting $p \to \infty$, and then taking the supremum over all finite subsets $\{\lambda_k\}$, we obtain $||x_n - x||_Y \leq \epsilon$ for all $n > n_o$. Thus Y is a Banach subspace of X.

If $U \in B(X)$ commutes with A, it commutes also with $R(\lambda; A)$ for each $\lambda > a$. Therefore, for $x \in Y$,

$$||Ux||_Y = \sup_{\lambda_k > a} ||U\Pi_k(\lambda_k - a)R(\lambda_k; A)x|| \leq ||U||_{B(X)}.||x||_Y < \infty,$$

and so Y is U-invariant and $||U||_{B(Y)} \leq ||U||_{B(X)}$. ||||

1.22. DEFINITION. The **Hille-Yosida space** Z for A is the closure of $D(A_Y)$ in Y.

The terminology is motivated by the following

1.23. THEOREM. Let A be an unbounded operator with $(a, \infty) \subset \rho(A)$ for some real a. Let Z be the Hille-Yosida space for A. Then A_Z, the part of A in Z, generates a C_o-semigroup $T(.)$ in Z that satisfies $||T(t)||_{B(Z)} \leq e^{at}$.

Moreover, Z is **maximal** in the following sense: if $W = (W, ||.||_W)$ is a Banach subspace of X such that A_W generates a C_o-semigroup in W with the above exponential growth, then W is a Banach subspace of Z.

PROOF. Since $R(\lambda; A)$ commutes with A for each $\lambda > a$, the linear manifold Y is $R(\lambda; A)$-invariant and $||R(\lambda; A)|Y||_{B(Y)} \leq ||R(\lambda; A)||_{B(X)} < \infty$ by Lemma 1.21. The identities

$$(\lambda I - A)R(\lambda; A)y = y \qquad (y \in Y)$$

$$R(\lambda; A)(\lambda I - A)y = y \qquad (y \in D(A_Y))$$

show then that $R(\lambda; A)|Y = R(\lambda; A_Y)$ for all $\lambda > a$.

If $y \in D(A_Y)$, then $y, Ay \in Y$, so that $R(\lambda; A)y(\in D(A)) \in Y$ and $AR(\lambda; A)y = \lambda R(\lambda; A)y - y \in Y$, that is, $R(\lambda; A)D(A_Y) \subset D(A_Y)$. Since $R(\lambda; A)|Y \in B(Y)$, it follows that Z is $R(\lambda; A)$- invariant, and

$$||R(\lambda; A)|Z||_{B(Z)} \leq ||R(\lambda; A)|Y||_{B(Y)} < \infty. \tag{1}$$

The above identities show then that

$$R(\lambda; A_Z) = R(\lambda; A)|Z \qquad (\lambda \in (a, \infty)). \tag{2}$$

In particuBlar, A_Z is closed.

For all $y \in Y$ and all finite sets $\{\lambda_k\} \subset (a, \infty)$,

$$||\Pi_k(\lambda_k - a)R(\lambda_k; A_Y)y||_Y = ||\Pi_k(\lambda_k - a)R(\lambda_k; A)y||_Y$$

$$= \sup_{\mu_j > a} ||\Pi_j(\mu_j - a)R(\mu_j; A)\Pi_k(\lambda_k - a)R(\lambda_k; A)y||$$

$$\leq \sup_{\nu_r > a} ||\Pi_r(\nu_r - a)R(\nu_r; A)y|| = ||y||_Y.$$

Therefore

$$||\Pi_k(\lambda_k - a)R(\lambda_k; A_Y)||_{B(Y)} \leq 1 \tag{3}$$

for any finite set $\{\lambda_k\} \subset (a, \infty)$, and the same is true with Y replaced by Z. In particular, taking singleton subsets of (a, ∞), we have

$$||R(\lambda; A_Y)||_{B(Y)} \leq \frac{1}{\lambda - a} \qquad (\lambda > a).$$

Therefore, for all $z \in D(A_Y)$,

$$||\lambda R(\lambda; A)z - z||_Y = ||R(\lambda; A)Az||_Y \leq ||R(\lambda; A_Y)||_{B(Y)}.||Az||_Y$$

$$\leq \frac{||Az||_Y}{\lambda - a} \to 0$$

as $\lambda \to \infty$, since $Az \in Y$. Thus $\lambda R(\lambda; A)z \to z$ in Y for all $z \in D(A_Y)$. For $z \in Z$ arbitrary, if $\epsilon > 0$ is given, there exists $z_o \in D(A_Y)$ such that $||z - z_o||_Y < \epsilon$, since $D(A_Y)$ is dense in Z by Definition 1.22. Then

$$||\lambda R(\lambda; A_Z)z - z||_Y \leq ||(\lambda R(\lambda; A_Z) - I)(z - z_o)||_Y + ||\lambda R(\lambda; A)z_o - z_o||_Y$$

$$\leq (\frac{\lambda}{\lambda - a} + 1)\epsilon + ||\lambda R(\lambda; A)z_o - z_o||_Y \to 2\epsilon$$

as $\lambda \to \infty$. Hence as $\lambda \to \infty$,

$$\lambda R(\lambda; A_Z)z(\in D(A_Z)) \to z$$

in the $||.||_Y$-norm, and so $D(A_Z)$ is dense in Z.

In conclusion, A_Z is a densely defined closed operator in Z, which satisfies in Z the condition (**) with $M = 1$. By the Hille-Yosida theorem (1.17), it follows that A_Z generates in Z a C_o-semigroup $T(.)$ satisfying $||T(t)||_Z \leq e^{at}$ for all $t \geq 0$.

On the other hand, if W is as in the statement of the theorem, then for any $w \in W$, it follows from (**) (with $M = 1$) in $B(W)$ that

$$||w||_Y \leq \sup_{\lambda_k > a} ||\Pi_k(\lambda_k - a)R(\lambda_k; A)||_{B(W)}||w||_W \leq ||w||_W.$$

21

Therefore W is a Banach subspace of Y. In particular $D(A_W) \subset D(A_Y)$. Since A_W generates a C_o-semigroup in W,

$$W = W\text{-closure}(D(A_W)) \subset W\text{-closure}(D(A_Y))$$

$$\subset Y\text{-closure}(D(A_Y)) := Z,$$

and we conclude that W is a Banach subspace of Z. ||||

Note in particular the case $a = 0$: if $(0, \infty) \subset \rho(A)$, the Hille-Yosida space for A is a maximal Banach subspace such that the part of A in it generates a C_o-semigroup of contractions in it.

C. DISSIPATIVITY

A useful characterization of generators that avoids resolvents depends on the numerical range.

1.24. DEFINITION. Let A be an arbitrary (usually unbounded) operator on the Banach space X. Its **numerical range** is the set

$$\nu(A) = \{x^*Ax; x \in D(A), x^* \in X^*, ||x|| = ||x^*|| = x^*x = 1\}.$$

Given $x \in D(A)$ with $||x|| = 1$, define x^* on $\mathbb{C}x$ by $x^*(\lambda x) = \lambda$ for $\lambda \in \mathbb{C}$. Then $||x^*|| = x^*x = 1$, and x^* extends to a unit vector in X^* by the Hahn-Banach theorem. This shows that $\nu(A)$ is not empty.

1.25. DEFINITION. The operator A is **dissipative** if $\Re\nu(A) \leq 0$.

1.26. THEOREM. If A generates a C_o-semigroup of contractions, then it is closed, densely defined, dissipative, and $\lambda I - A$ is surjective for all $\lambda > 0$.

Conversely, if A is closed, densely defined, dissipative, and $\lambda I - A$ is surjective for all $\lambda > \lambda_o$ (for some $\lambda_o \geq 0$), then A generates a C_o-semigroup of contractions.

PROOF. *Necessity.* Suppose A generates the C_o-semigroup of contractions $T(.)$, and let $x \in X$ and $x^* \in X^*$ be unit vectors such that $x^*x = 1$. For $h > 0$, $|x^*T(h)x| \leq ||x^*||.||T(h)||.||x|| \leq 1$, and therefore

$$\Re x^*[h^{-1}(T(h)x - x)] = h^{-1}[\Re(x^*T(h)x) - 1] \leq 0.$$

For $x \in D(A)$, letting $h \to 0$, we get $\Re x^*Ax \leq 0$, so that A is dissipative. It is closed and densely defined by Theorem 1.2. By Corollary 1.18, $(0, \infty) \subset \rho(A)$, so that, in particular, $\lambda I - A$ is surjective for all $\lambda > 0$.

Sufficiency. For all unit vectors $x \in D(A), x^* \in X^*$ such that $x^*x = 1$, and for all $\lambda > 0$, we have

$$||(\lambda I - A)x||^2 \geq |x^*(\lambda I - A)x|^2 = |\lambda - x^*Ax|^2 \tag{1}$$

$$= \lambda^2 - 2\lambda\Re(x^*Ax) + |x^*Ax|^2 \geq \lambda^2$$

because $\Re(x^*Ax) \leq 0$. Therefore $\lambda I - A$ is one-to-one (for all $\lambda > 0$) and onto X (by hypothesis) for all $\lambda > \lambda_o$. Thus $\lambda \in \rho(A)$, and $||\lambda R(\lambda; A)|| \leq 1$ for all $\lambda > \lambda_o$. This proves that A generates a C_o-contraction semigroup, by Corollary 1.18 (it is clear from the proof of Theorem 1.17 that for the sufficiency part, the growth condition on the resolvents is needed for large λ only).|||||

Note that in (1), we used only *some* unit vector x^* with the needed properties. This allows the following weakening of the hypothesis in the sufficiency part of the theorem.

1.27. THEOREM. Let A be a closable densely defined operator such that $\lambda_o I - A$ has dense range for some $\lambda_o > 0$. Suppose that for each $x \in D(A)$, there exists a unit vector $x^* \in X^*$ such that $x^*x = ||x||$ and $\Re(x^*Ax) \leq 0$. Then the closure A^- of A generates a C_o- semigroup of contractions.

PROOF. As in (1), $||(\lambda I - A)x|| \geq \lambda||x||$ for all $\lambda > 0$ and $x \in D(A)$. Let $x \in D(A^-)$, and let then $x_n \in D(A)$ be such that $x_n \to x$ and $Ax_n \to A^-x$. Letting $n \to \infty$ in the inequalities $||(\lambda I - A)x_n|| \geq \lambda||x_n||$, we obtain

$$||(\lambda I - A^-)x|| \geq \lambda||x|| \qquad (\lambda > 0; x \in D(A^-)). \qquad (2)$$

In particular, $\lambda I - A^-$ is one-to-one for all $\lambda > 0$. We claim that $\lambda_o I - A^-$ is onto X. Indeed, for any $y \in X$, there exist by hypothesis $x_n \in D(A)$ such that $(\lambda_o I - A)x_n \to y$. Then by (2),

$$||x_n - x_m|| \leq \lambda_o^{-1}||(\lambda_o I - A)(x_n - x_m)|| \to 0,$$

so $x_n \to x$, and necessarily $x \in D(A^-)$ and $(\lambda_o I - A^-)x = y$.

Thus $\lambda_o \in \rho(A^-)$ and $||R(\lambda_o; A^-)|| \leq 1/\lambda_o$. By Theorem 1.11, $d(\lambda_o, \sigma(A^-)) \geq \frac{1}{||R(\lambda_o; A^-)||} \geq \lambda_o$. Therefore $(0, 2\lambda_o) \subset \rho(A^-)$. Inductively, one obtains that $(0, 2^n\lambda_o) \subset \rho(A^-)$ for all n, and so $(0, \infty) \subset \rho(A^-)$ and $\lambda R(\lambda; A^-)$ are contractions for all $\lambda > 0$, by (2). The result follows now from Corollary 1.18.|||||

The criterion of Theorem 1.26 is effective for certain types of perturbations of generators.

1.28. DEFINITION. Let A, B be (usually unbounded) operators. One says that B is **A-bounded** if $D(A) \subset D(B)$ and there exist $a, b \geq 0$ such that

$$||Bx|| \leq a||Ax|| + b||x|| \qquad (x \in D(A)).$$

The infimum of all a as above is called the **A-bound** of B.

For example, any $B \in B(X)$ is A-bounded with A-bound equal to 0.

1.29. LEMMA. If A is closed and B is A-bounded with A-bound $a < 1$, then $A + B$ (with domain $D(A)$) is closed.

PROOF. Note first that the A-boundedness of B means that $B \in B([D(A)], X)$ (recall that $[D(A)]$ is normed by the graph-norm for A).

Let $x_n \in D(A), x_n \to x$, and $(A + B)x_n \to y$. Then

$$||Ax_n - Ax_m|| = ||(A + B)x_n - (A + B)x_m - B(x_n - x_m)||$$

$$\leq ||(A + B)x_n - (A + B)x_m|| + a||Ax_n - Ax_m|| + b||x_n - x_m||.$$

Hence

$$(1 - a)||Ax_n - Ax_m|| \leq ||(A + B)x_n - (A + B)x_m|| + b||x_n - x_m|| \to_{n,m \to \infty} 0.$$

Since $a < 1$, $\{Ax_n\}$ is Cauchy, and since A is closed, it follows that $x \in D(A)(= D(A + B))$ and $Ax_n \to Ax$. Since B is continuous on $[D(A)]$, also $Bx_n \to Bx$, and therefore $(A + B)x_n \to (A + B)x.||||$

We now have the following perturbation theorem.

1.30. THEOREM. Let A generate a C_o-semigroup of contractions, and let B be dissipative and A-bounded with A-bound $a < 1$. Then $A + B$ generates a C_o-semigroup of contractions.

PROOF. Since A generates a C_o-contraction semigroup, it is dissipative (1.26). Also B is dissipative with $D(A) \subset D(B)$. Therefore $A + B$ is dissipative, because for all $x \in D(A + B) = D(A)$ and $x^* \in X^*$ with $||x|| = ||x^*|| = x^*x = 1$, $\Re[x^*(A + B)x] = \Re[x^*Ax] + \Re[x^*Bx] \leq 0$. By Lemma 1.29, $A + B$ is closed, and it is densely defined ($D(A)$ is dense by Theorem 1.2). By Theorem 1.26, it remains to show that $\lambda I - (A + B)$ is surjective for all $\lambda > \lambda_o$ (for some $\lambda_o \geq 0$). We have for all $\lambda > 0$

$$ran (\lambda I - A - B) = [(\lambda I - A) - B]R(\lambda; A)X = [I - BR(\lambda; A)]X. \qquad (*)$$

However, for all $x \in X$,

$$||BR(\lambda; A)x|| \leq a||AR(\lambda; A)x|| + b||R(\lambda; A)x||$$

$$\leq a||\lambda R(\lambda; A)||.||x|| + a||x|| + \frac{b}{\lambda}||\lambda R(\lambda; A)||.||x||$$

$$\leq (2a + \frac{b}{\lambda})||x||$$

since $\lambda R(\lambda; A)$ are contractions, by Corollary 1.18.

In case $a < 1/2$, $2a + \frac{b}{\lambda} < 1$ for $\lambda > \lambda_o$, and therefore $||BR(\lambda; A)|| < 1$, hence $I - BR(\lambda; A)$ is invertible in $B(X)$, and so $\lambda I - (A + B)$ is surjective for $\lambda > \lambda_o$, by (*).

Consider now the general case $a < 1$. Let $t_1 = \frac{1-a}{2}$. For any $s \in [0, 1]$ and $x \in D(A)$,

$$(1 - as)||Bx|| = ||Bx|| - as||Bx|| \le (a||Ax|| + b||x||) - as||Bx||$$

$$= a(||Ax|| - s||Bx||) + b||x|| \le a||(A + sB)x|| + b||x||.$$

Therefore, for $t \in [0, t_1]$,

$$||tBx|| \le \frac{1 - as}{2}||Bx|| \le \frac{a}{2}||(A + sB)x|| + \frac{b}{2}||x||,$$

so that tB has an $(A + sB)$-bound $< \frac{1}{2}$. By the preceding case, if $A + sB$ generates a C_o-contraction semigroup for some $s \in [0, 1]$, so does $A + sB + tB$ for all $t \in [0, t_1]$. Starting with $s = 0$ (for which $A + sB = A$ is a generator by hypothesis), we get that $A + tB$ is a generator for all $t \in [0, t_1]$, hence $A + t_1 B + tB$ is a generator for all such t, etc... . Let n be the first integer such that $nt_1 \ge 1$. A last application of the above argument with $s = (n - 1)t_1 < 1$ gives that $A + tB$ is the generator of a C_o-contraction semigroup for all $t \in [0, nt_1]$, so that, in particular, $A + B$ is a generator.||||

D. THE TROTTER-KATO CONVERGENCE THEOREM

We consider next a one-parameter family $T_s(.)$ of C_o-semigroups ($s \in [0,c)$), with generators A_s; let us write $T(.) = T_0(.)$ and $A = A_0$.

BASIC HYPOTHESIS: there exist $M > 0$ and $a \geq 0$ such that

$$\|T_s(t)\| \leq M e^{at} \quad (t \geq 0, s \in [0,c)). \tag{1}$$

This implies (cf. (*) preceding 1.16)

$$\|R(\lambda; A_s)\| \leq \frac{M}{\lambda - a} \tag{2}$$

for all $\lambda > a$ and $s \in [0,c)$.

We define the following convergence properties:

1.31. DEFINITION. 1. **Generators graph convergence** on a **core** D_o for A: for each $x \in D_o$, there exists $x_s \in D(A_s)$ such that $[x_s, A_s x_s] \to [x, Ax]$ in X^2 when $s \to 0$.
   2. **Resolvents strong convergence**: for each $\lambda > a$, $R(\lambda; A_s) \to R(\lambda; A)$ in the strong operator topology (when $s \to 0$).
   3. **Semigroups strong uniform convergence on compacta**: for each $x \in X$, $T_s(t)x \to T(t)x$ in X, uniformly on compact t-intervals (when $s \to 0$).

1.32. THEOREM (Trotter-Kato). The convergence properties in Definition 1.31 are equivalent.

PROOF. In the following, the numbers 1,2,3 refer to the three types of convergence formulated in Definition 1.31.

Denoting $y_s = (\lambda I - A_s)x_s$ and $y = (\lambda I - A)x$ for all $x \in D_o$, we see that 1. is equivalent to

$$1'. \qquad [y_s, R(\lambda; A_s)y_s] \to [y, R(\lambda; A)y]$$

for all $y \in (\lambda I - A)D_o$ (for suitable y_s, and for all $\lambda > a$).
   By (2), Property 1'. is equivalent to

$$1''. \qquad [y_s, R(\lambda; A_s)y] \to [y, R(\lambda; A)y]$$

for all $y \in (\lambda I - A)D_o$ (for suitable y_s, and for all $\lambda > a$).

1. implies 2. Since we are assuming 1''., we have in particular

$$R(\lambda; A_s)y \to R(\lambda; A)y \qquad (3)$$

for all $y \in (\lambda I - A)D_o$. Since D_o is a core for A, for each $x \in D(A)$, there exist $x_n \in D_o$, $n = 1, 2, ...$, such that $[x_n, Ax_n] \to [x, Ax]$, hence $[x_n, (\lambda I - A)x_n] \to [x, (\lambda I - A)x]$. In particular, $X = (\lambda I - A)D(A) = [(\lambda I - A)D_o]^-$.

Thus (3) is valid for y in a dense subspace of X, hence for all $y \in X$, by (2). This is Property 2.

2. implies 1. Given $x \in D_o$ and $\lambda > a_,$, choose $x_s = R(\lambda; A_s)(\lambda I - A)x$. Correspondingly, we have $y = (\lambda I - A)x$ and $y_s = (\lambda I - A)x = y$, so that by 2.

$$[y_s, R(\lambda; A_s)y] = [y, R(\lambda; A_s)y] \to [y, R(\lambda; A)y].$$

Thus 1''. (and so 1.) is satisfied.

For the implication 2. $\Longrightarrow$ 3., we need the following

LEMMA. Let A, B generate the C_o-semigroups $T(.)$ and $V(.)$ respectively, both $O(e^{at})$ for some $a \geq 0$. Then for Re $\lambda > a, t \geq 0$ and $x \in X$,

$$R(\lambda; B)[V(t) - T(t)]R(\lambda; A)x = \int_0^t V(t - s)[R(\lambda; B) - R(\lambda; A)]T(s)x ds.$$

PROOF (of Lemma). For λ, t as above and $0 \leq s \leq t$,

$$\frac{d}{ds}V(t - s)R(\lambda; B)T(s)R(\lambda; A)x = V(t - s)(-B)R(\lambda; B)T(s)R(\lambda; A)x$$

$$+V(t - s)R(\lambda; B)T(s)AR(\lambda; A)x$$

$$= V(t - s)[T(s)R(\lambda; A)x - \lambda R(\lambda; B)T(s)R(\lambda; A)x]$$

$$+V(t - s)R(\lambda; B)T(s)[\lambda R(\lambda; A)x - x]$$

$$= V(t - s)[R(\lambda; A) - R(\lambda; B)]T(s)x.$$

Integrating with respect to s from 0 to t, we obtain the formula in the lemma.

2. $\Longrightarrow$ 3. For all $y \in X$, write

$$[T_s(t) - T(t)]R(\lambda; A)y = R(\lambda; A_s)[T_s(t) - T(t)]y$$

$$+T_s(t)[R(\lambda; A) - R(\lambda; A_s)]y$$

$$+[R(\lambda; A_s) - R(\lambda; A)]T(t)y = I + II + III \qquad (s, t \geq 0, \lambda > a).$$

We estimate I for $y = R(\lambda; A)x$, using the lemma. Thus for $0 \leq t \leq \tau$,

$$||I|| \leq \int_0^\tau M e^{a(\tau-u)} ||[R(\lambda; A_s) - R(\lambda; A)]T(u)x|| du.$$

The right hand side converges to zero when $s \to 0$, by 2.; therefore $I \to 0$ uniformly on every compact t-interval. Since $R(\lambda; A)X = D(A)$ is dense in X, and the operators in I are uniformly bounded with respect to s (and with respect to t in compacta (cf. (1) and (2)), it follows that I converges to zero for all $y \in X$, uniformly on compact t-intervals (when $s \to 0$).

By (1) and 2., for $0 \leq t \leq \tau$,

$$||II|| \leq M e^{a\tau} ||R(\lambda; A)y - R(\lambda; A_s)y|| \to_{s \to 0} 0,$$

so that $II \to 0$ uniformly on compact t-intervals.

For $y \in D(A)$, we may write

$$T(t)y = y + \int_0^t T(r)Ay \, dr,$$

and therefore, for $0 \leq t \leq \tau$,

$$||III|| \leq ||[R(\lambda; A_s) - R(\lambda; A)]y||$$

$$+ \int_0^\tau ||[R(\lambda; A_s) - R(\lambda; A)]T(r)Ay|| dr.$$

By 2., the first term on the right converges to 0 as $s \to 0$. The integrand on the right converges pointwise to 0, and is bounded by $\frac{2M^2}{\lambda-a} e^{a\tau} ||Ay||$ in the interval $[0, \tau]$. By Lebesgue's Dominated Convergence theorem, the integral converges to 0, and therefore $III \to 0$ uniformly on compact t-intervals. Since $D(A)$ is dense in X and the operators appearing in III are uniformly bounded (with respect to $s \in [0, c)$ and t in compacta; cf. (1) and (2)), the above conclusion is valid for all $y \in X$.

We thus obtained that $T_s(t)x \to T(t)x$ (when $s \to 0$), uniformly on compact t-intervals, for all $x \in R(\lambda; A)X = D(A)$ ($\lambda > a$ fixed); since $D(A)$ is dense in X and $||T_s(t) - T(t)|| \leq 2M e^{at}$ are uniformly bounded for all $s \geq 0$ and for all t in compacta, Property 3. follows.

3. $\Rightarrow$ 2. For $\lambda > a$ and $x \in X$, $R(\lambda; A_s)x = \int_0^\infty e^{-\lambda t} T_s(t)x \, dt$. When $s \to 0$, the integrand converges pointwise to its value at $s = 0$ (by 3.) and is norm-dominated

by $Me^{-(\lambda-a)t}||x|| \in L^1(0,\infty)$. By Lebesgue's Dominated Convergence theorem, the integral converges to its value at $s = 0$, which is precisely $R(\lambda; A)x$. ||||

1.33. COROLLARY. (Same "basic hypothesis" (1)). Suppose that for each x in a core D_o for A, there exists $s_o \in (0,c)$ such that $x \in D(A_s)$ for all $s \in (0, s_o)$ and $A_s x \rightarrow Ax$ when $s \rightarrow 0+$. Then $T_s(t)x \rightarrow T(t)x$ for all $x \in X$, uniformly on compact t-intervals (when $s \rightarrow 0+$).

PROOF. Property 1. is satisfied with $x_s = x$ for $s \in (0, s_o)$. Therefore Property 3. holds, by Theorem 1.32. ||||

1.34. COROLLARY. Let $T(.)$ be a C_o-semigroup, and denote $A_s = s^{-1}[T(s)-I]$. Then

$$T(t) = \lim_{s \rightarrow 0+} e^{tA_s}$$

strongly and uniformly on compact t-intervals.

PROOF. Let $r > \omega$, and choose $a = re^r$. Let $T_s(t) = e^{tA_s}$ for $t \geq 0, s \in (0,1)$, and $T_0(.) = T(.)$. For $M = M_r$, we have $||T(t)|| \leq Me^{rt} \leq Me^{at}$ and

$$||T_s(t)|| = e^{-t/s}||e^{(t/s)T(s)}|| = e^{-t/s}||\Sigma_{n \geq 0}(t/s)^n T(ns)/n!||$$

$$\leq Me^{-t/s}\Sigma(t/s)^n e^{nsr}/n! = M \exp[ts^{-1}(e^{sr} - 1)] \leq Me^{at}$$

for all $t \geq 0$ and $s \in (0,1)$.

Thus the "basic hypothesis" (1) is satisfied (with $c = 1$), and since $A_s x \rightarrow Ax$ for all $x \in D(A)$ (by definition of A), our corollary follows from Corollary 1.33.||||

E. EXPONENTIAL FORMULAS

A useful application of Corollary 1.33 is the following

1.35. THEOREM. Let A generate a C_o-contraction semigroup $T(.)$, and let F be *any* contraction-valued function on $[0, \infty)$ such that $F(0) = I$ and the right derivative at zero of $F(.)x$ coincides with Ax for all x in a core D_o for A. Then $T(.)$ is the strong limit of $F(t/n)^n$ (as $n \to \infty$), uniformly on compact t-intervals.

PROOF. We need the following

LEMMA. Let C be a contraction on X. Then $e^{t(C-I)}$ is a uniformly continuous contraction semigroup, and

$$||e^{n(C-I)}x - C^n x|| \leq n^{1/2}||(C - I)x||$$

for all $x \in X$ and $n = 1, 2, \ldots$.

PROOF (of Lemma).

$$||e^{t(C-I)}|| = e^{-t}||e^{tC}|| \leq e^{-t}e^{t||C||} \leq 1.$$

$$||e^{n(C-I)}x - C^n x|| \leq e^{-n}||\Sigma_{k \geq 0}(n^k/k!)[C^k x - C^n x]||$$

$$= e^{-n}||\Sigma_{0 \leq k \leq n}(n^k/k!)C^k(x - C^{n-k}x) + C^n\Sigma_{k > n}(n^k/k!)(C^{k-n}x - x)||$$

$$\leq e^{-n}\Sigma_{k \geq 0}(n^k/k!)||C^{|k-n|}x - x||.$$

Since C is a contraction,

$$||C^m x - x|| = ||(C^{m-1} + \ldots + I)(C - I)x|| \leq m||Cx - x||,$$

and therefore the last expression is

$$\leq [\Sigma_{k \geq 0}e^{-n}(n^k/k!)|k - n|].||Cx - x||.$$

The term in square brackets is the expectation of $|K - n|$, where K is a Poisson random variable with parameter n. By Schwarz' inequality, since K has expectation and variance n, we have $E(|K - n|) \leq [E(K - n)^2]^{1/2} := \sigma(K) = n^{1/2}.||||$

31

Back to the proof of the theorem, consider the bounded operators

$$A_n = (t/n)^{-1}[F(t/n) - I]$$

for t fixed. By hypothesis, $A_n x \to A x$ for all $x \in D_o$. For all unit vectors $x \in X$ and $x^* \in X^*$ such that $x^* x = 1$, we have $\Re(x^* A_n x) = (n/t)[\Re(x^* F(t/n)x) - 1] \leq 0$ because $|x^* F(.)x| \leq ||x^*||.||F(.)||.||x|| \leq 1$. Thus A_n is dissipative, and so, by 1.26, e^{sA_n} are contraction semigroups satisfying trivially the "basic hypothesis" (with $M = 1$ and $a = 0$). By Corollary 1.33,

$$e^{sA_n} \to T(s) \tag{1}$$

strongly and uniformly on compact s-intervals, when $n \to \infty$.

However, by the lemma with $C = F(t/n)$,

$$||e^{tA_n}x - F(t/n)^n x|| \leq n^{1/2}||[F(t/n) - I]x|| = \frac{t}{n^{1/2}}||A_n x|| \to 0 \tag{2}$$

as $n \to \infty$, for all $x \in D_o$. Since D_o is dense in X (as a core) and $||e^{tA_n} - F(t/n)^n|| \leq 2$ for all n because both operators are contractions, it follows that (2) is valid for all $x \in X$. By (1) and (2), the theorem follows.||||

As a first application of Theorem 1.35, we obtain the following "exponential formula".

1.36. THEOREM. Let $T(.)$ be a C_o-semigroup, with generator A. Then for all $t > 0$,

$$T(t) = \lim_n [\frac{n}{t} R(\frac{n}{t}; A)]^n$$

in the strong operator topology (as $n \to \infty$).

PROOF. We consider first the case when $||T(t)|| \leq e^{at}$ for all $t \geq 0$ (with some $a \geq 0$).

Let $S(t) := e^{-at}T(t)$. This is a C_o-semigroup of contractions, with generator $A - aI$.

We set

$$F(s) := (s^{-1} - a)R(s^{-1}; A) = (s^{-1} - a)R(s^{-1} - a; A - aI) \qquad (s \in (0, 1/a)),$$

and $F(0) = I$.

By Corollary 1.18, $F(.)$ is contraction-valued. By Lemma 1.16,

$$s^{-1}[F(s)x - x] = A_{1/s}x - as^{-1}R(s^{-1}; A)x \to (A - aI)x$$

for all $x \in D(A) = D(A - aI)$, as $s \to 0+$. We may then apply Theorem 1.35; thus, in the strong operator topology and uniformly in compact t-intervals,

$$S(t) = \lim_n F(t/n)^n,$$

and therefore

$$T(t) = e^{at} S(t) = \lim_n [(1 - \frac{at}{n})^{-1} F(t/n)]^n$$

$$= \lim_n [\frac{n}{n - at} \frac{n - at}{t} R(\frac{n}{t}; A)]^n = \lim_n [\frac{n}{t} R(\frac{n}{t}; A)]^n.$$

General case. Fix $a > \omega$. We have $\|T(t)\| \leq M e^{at}$ for all $t \geq 0$ (with a suitable M depending only on a). Renorm X by $|x| := \sup_{t \geq 0} e^{-at} \|T(t)x\|$. Then $\|x\| \leq |x| \leq M\|x\|$, i.e., the norms are equivalent, and

$$|T(t)x| = \sup_{s \geq 0} e^{-as} \|T(s + t)x\| \leq e^{at} \sup_{u \geq 0} e^{-au} \|T(u)x\| = e^{at}|x|.$$

We may then apply the preceding case to the semigroup $T(.)$ on the space $(X, |.|)$, yielding the result with respect to the $|.|$-norm, hence also with respect to the given norm (since the two norms are equivalent).$\|\|\|$

Another application of Theorem 1.35 is the following

1.37. THEOREM (**Trotter's Product Formula**). Let A, B, C generate contraction C_o-semigroups $S(.), T(.), U(.)$ respectively, and suppose that $C = A + B$ on a core D_o for C. Then for all $t \geq 0$,

$$U(t) = \lim_n [S(t/n) T(t/n)]^n$$

strongly.

PROOF. Take $F(t) = S(t)T(t)$ in Theorem 1.35. For $x \in D_o$ and $t > 0$,

$$t^{-1}[F(t)x - x] = S(t)t^{-1}[T(t)x - x] + t^{-1}[T(t)x - x] \to Bx + Ax = Cx,$$

and the conclusion follows from Theorem 1.35.$\|\|\|$

F. THE HILLE-PHILLIPS PERTURBATION THEOREM

The next theorem is concerned with perturbations $A + B$ of the generator A by closed operators B in a suitable class.

FIXED HYPOTHESIS H_1. Let A generate a C_o-semigroup $T(.)$ and let B be a closed operator such that $T(t)X \subset D(B)$ for all $t > 0$.

By the Closed Graph Theorem, $BT(t) \in B(X)$ for all $t > 0$. Hence, if $t, t+h > 0$, $BT(t+h)x - BT(t)x = [BT(t)][T(h)x - x] \to 0$ when $h \to 0$, showing that $BT(.)$ is strongly right continuous on $(0, \infty)$. It follows that $||BT(.)||$ is bounded on compact intervals, and we deduce that $BT(.)$ is strongly continuous for $t > 0$, as in the proof of Theorem 1.1. Thus $||BT(.)||$ is measurable on $t > 0$ (cf. proof of Theorem 1.2).
Also, for any $t > \epsilon > 0$,

$$\frac{\log ||BT(t)||}{t} \leq \frac{\log ||BT(\epsilon)||}{t} + (1 - \frac{\epsilon}{t})\frac{\log ||T(t - \epsilon)||}{t - \epsilon} \to \omega,$$

so that

$$\limsup_{t \to \infty} \frac{\log ||BT(t)||}{t} \leq \omega.$$

Therefore, for any $a > \omega$, there exists a constant $M_a > 0$ such that $||BT(t)|| \leq M_a e^{at}$.

The non-negative measurable function $||BT(.)||$ has an integral over $[0, 1]$ (that could be infinite). We assume

FIXED HYPOTHESIS H_2. $\int_0^1 ||BT(t)||dt < \infty$.

Note that any $B \in B(X)$ satisfies H_1 and H_2 trivially.

1.38. THEOREM. Let A, B satisfy the hypothesis H_1 and H_2. Then $A + B$ (with domain $D(A)$) generates a C_o-semigroup.

More details about the structure of the semigroup generated by the perturbation $A + B$ will be obtained in Lemma 3. and in (**) below.

LEMMA 1. (Hypothesis H_1, H_2). For $\Re\lambda > \omega$, $D(A)(= R(\lambda; A)X) \subset D(B)$ and

$$BR(\lambda; A)x = \int_0^\infty e^{-\lambda t} BT(t)x dt$$

for all $x \in X$, where the Laplace integral above converges absolutely.

PROOF. The Riemann sums S_n for the integral $\int_a^b$ (with $0 < a < b < \infty$) approximating the Laplace integral of $T(.)x$ are in $D(B)$ by H_1, converge to $\int_a^b$, and by linearity of B and continuity of $BT(.)x$ in $[a, b]$, BS_n converge to $\int_a^b e^{-\lambda t} BT(t)x \, dt$. Since B is closed, it follows that $\int_a^b \in D(B)$ and $B\int_a^b = \int_a^b e^{-\lambda t} BT(t)x \, dt$. Since

$$\int_0^\infty ||e^{-\lambda t} BT(t)|| dt \le e^{-\Re\lambda} \int_0^1 ||BT(t)|| dt + \int_1^\infty e^{-\Re\lambda t} ||BT(t)|| dt < \infty$$

by H_2 and the remarks following H_1, the Laplace integral of $BT(.)x$ converges absolutely in X to an element $L(\lambda)x \in X$. By Theorem 1.15, $\int_a^b e^{-\lambda t} T(t)x \, dt (\in D(B)) \to R(\lambda; A)x$ (when $a \to 0$ and $b \to \infty$), and we observed before that $B\int_a^b (...) \to L(\lambda)x$. Since B is closed, it follows that $R(\lambda; A)x \in D(B)$ and $BR(\lambda; A)x = L(\lambda)x$.||||

LEMMA 2. (Hypothesis H_1, H_2). There exists $r > \omega$ such that

$$q := \int_0^\infty e^{-rt} ||BT(t)|| dt < 1.$$

For $\Re\lambda > r$,
$$R(\lambda; A + B) = R(\lambda; A) \Sigma_{n \ge 0} [BR(\lambda; A)]^n, \tag{*}$$

where the series converges in $B(X)$.

In particular, $A + B$ is closed (and densely defined, since $D(A + B) = D(A)$, by Lemma 1.).

PROOF. Fix $c > \omega$. By H_1, for all $\lambda > c$, $e^{-\lambda t} ||BT(t)|| \le e^{-ct} ||BT(t)|| \in L^1(0, \infty)$, and $e^{-\lambda t} ||BT(t)|| \to 0$ as $\lambda \to \infty$. By Dominated Convergence, it follows that $\int_0^\infty e^{-\lambda t} ||BT(t)|| dt \to 0$ when $\lambda \to \infty$. We may then choose $r > c$ such that $q < 1$. Then, by Lemma 1., $||BR(\lambda; A)|| \le q < 1$ for $\Re\lambda > r$, and therefore the right hand side of (*) converges in $B(X)$ to an operator $K(\lambda)$ with range in $D(A) = D(A + B)$. We have for $\Re\lambda > r$

$$[\lambda I - (A + B)]K(\lambda) = (\lambda I - A)K(\lambda) - BK(\lambda)$$

$$= \Sigma_{n \ge 0} [BR(\lambda; A)]^n - \Sigma_{n \ge 0} [BR(\lambda; A)]^{n+1} = I.$$

On the other hand, for $x \in D(A)$,

$$K(\lambda)[\lambda I - (A + B)]x = R(\lambda; A)\{I + \Sigma_{n \ge 1} [BR(\lambda; A)]^n\}[(\lambda I - A) - B]x$$

35

$$= x + R(\lambda; A)\{-Bx + \Sigma_{n\geq 0}[BR(\lambda; A)]^n[BR(\lambda; A)][(\lambda I - A) - B]x\}$$
$$= x + R(\lambda; A)\{-Bx + \Sigma_{n\geq 0}[BR(\lambda; A)]^n Bx - \Sigma_{n\geq 1}[BR(\lambda; A)]^n Bx\} = x. |||||$$

The functions $f := ||T(.)||$ and $g := ||BT(.)||$ are both in L^1_{loc}, the class of locally integrable functions on $(0, \infty)$ (cf. remarks following H_1, together with H_2). The "Laplace" convolution

$$(u * v)(t) := \int_0^t u(t - s)v(s)ds \qquad u, v \in L^1_{loc}$$

defines a function in L^1_{loc}, and therefore the repeated convolutions

$$g^{(n)} = g * \ldots * g \quad \text{n times}$$

are in L^1_{loc}. We consider also the L^1_{loc}-functions $h^{(n)} = f * g^{(n)}$, and we set $h^{(0)} = f$.
The next lemma will justify the following inductive definition:

$$S_0(.) = T(.); \qquad S_n(t)x = \int_0^t T(t - s)BS_{n-1}(s)xds \qquad (n = 1, 2, \ldots).$$

LEMMA 3. For all $n = 0, 1, 2, \ldots$, $S_n(.)$ are well-defined bounded operators such that, for all $x \in X$,
  a. $S_n(.)x : [0, \infty) \to D(B)$ is continuous, and for r, q as in Lemma 2. and all $t \geq 0$,
$$||S_n(t)|| \leq h^{(n)}(t) \leq Me^{rt}q^n;$$

  b. $BS_n(.)x : (0, \infty) \to X$ is continuous and $||BS_n(t)|| \leq g^{(n+1)}(t)$.

PROOF. We prove the lemma by induction on n. The case $n = 0$ is trivial (see observations following H_1).
Assume then the lemma's claims are valid for some n. By b. for n, $S_{n+1}(.)$ is well-defined and

$$S_{n+1}(t)x = \lim \int_a^b T(t - s)BS_n(s)xds \qquad (*)$$

as $a \to 0+$ and $b \to t-$. The Riemann sums for each integral over $[a, b]$ are in $D(B)$ by H_1, and when B is applied to them, the new sums converge to $\int_a^b BT(t - s)BS_n(s)xds$, because the latter's integrand is continuous on $[a, b]$ by b. (for n). Since B is closed, it follows that each integral in $(*)$ belongs to $D(B)$, and $B\int_a^b(\ldots) = \int_a^b B(\ldots)$. The same type of argument with $a \to 0+$ and $b \to t-$ (using again the closeness of B) shows that $S_{n+1}(t)x \in D(B)$ and

$$BS_{n+1}(t)x = \int_0^t BT(t - s)BS_n(s)xds.$$

Since $T(.), BT(.)$ and $BS_n(.)$ are continuous on $(0, \infty)$ and majorized by L^1_{loc}-functions (using the induction hypothesis), the integral representations for $S_{n+1}(.)x$ and $BS_{n+1}(.)x$ imply their continuity on $(0, \infty)$ for each $x \in X$. The function $S_{n+1}(.)$ is even norm-continuous at 0, since

$$||S_{n+1}(t)|| \leq f * g^{(n+1)}(= h^{(n+1)}) \leq Me^{rt} \int_0^t g^{(n+1)}(s)ds \to 0$$

when $t \to 0+$, because the integrand is in L^1_{loc}. Note that $S_n(0) = 0$ for all $n \geq 1$.

The estimates in a. and b. for $n + 1$ are trivial consequences of the induction hypothesis. For example,

$$h^{(n+1)} = f * g^{(n+1)} = [f * g^{(n)}] * g = h^{(n)} * g,$$

so that

$$h^{(n+1)}(t) \leq Mq^n \int_0^t e^{r(t-s)} g(s)ds = Mq^n e^{rt} \int_0^t e^{-rs} g(s)ds \leq Me^{rt} q^{n+1}.||||$$

The exponential growth of $S_n(.)$ (as in a.) shows that its Laplace transform converges absolutely for $\Re\lambda > r$. We show next

LEMMA 4. For $\Re\lambda > r$ and $x \in X$,

$$R(\lambda; A)[BR(\lambda; A)]^n x = \int_0^\infty e^{-\lambda t} S_n(t)x dt \qquad n = 0, 1, 2, ...$$

PROOF. The case $n = 0$ is verified by Theorem 1.15. Assuming the lemma for n, we have by Theorem 1.15,

$$R(\lambda; A)[BR(\lambda; A)]^{n+1} x = \int_0^\infty e^{-\lambda t} T(t)[BR(\lambda; A)]^{n+1} x dt$$

$$= \int_0^\infty e^{-\lambda t} T(t)B.\{R(\lambda; A)[BR(\lambda; A)]^n x\} dt$$

$$= \int_0^\infty e^{-\lambda t} T(t)B \int_0^\infty e^{-\lambda s} S_n(s)x ds dt,$$

where we used the induction hypothesis. Since B is closed, the argument we used before (cf. Lemmas 1. and 3.) allows us to move B inside the inner integral, and then do the same with the bounded operator $T(t)$. We then obtain the repeated integral

$$\int_0^\infty \int_0^\infty e^{-\lambda(t+s)} T(t) B S_n(s) x \, ds \, dt = \int_0^\infty e^{-\lambda u} \int_0^u T(u-s) B S_n(s) x \, ds \, du$$

$$= \int_0^\infty e^{-\lambda u} S_{n+1}(u) x \, du,$$

where the interchange of integration order is justified by absolute convergence.||||

We state now a simple characterization of generators as our final

LEMMA 5. An operator A on the Banach space X is the generator of a C_0-semigroup if and only if it is closed, densely defined, and for all $\lambda > a$ and $x \in X$,

$$R(\lambda; A) x = \int_0^\infty e^{-\lambda t} S(t) x \, dt,$$

where $S(.) : [0, \infty) \to B(X)$ is strongly continuous and $||S(t)|| \le M e^{at}$. In that case, $S(.)$ is the semigroup generated by A.

PROOF. Necessity follows from Theorems 1.1, 1.2, and 1.15.
Sufficiency. The series expansion for the resolvent obtained in the proof of Theorem 1.11 shows that

$$(-1)^n R(\lambda; A)^{(n)} = n! R(\lambda; A)^{n+1}.$$

The exponential growth of $||S(.)||$ allows us (using a Dominated Convergence argument) to differentiate the Laplace transform of $S(.)x$ under the integral sign, yielding inductively to the formula

$$[R(\lambda; A) x]^{(n)} = \int_0^\infty (-t)^n e^{-\lambda t} S(t) x \, dt.$$

Therefore

$$||R(\lambda; A)^n x|| = \frac{1}{(n-1)!} ||[R(\lambda; A) x]^{(n-1)}|| \le \frac{1}{(n-1)!} \int_0^\infty t^{n-1} e^{-\lambda t} ||S(t)|| dt . ||x||$$

$$\le \frac{M||x||}{(n-1)!} \int_0^\infty e^{-(\lambda-a)t} t^{n-1} dt = \frac{M}{(\lambda-a)^n} ||x||,$$

and the Hille-Yosida Theorem applies. If $T(.)$ is the C_0-semigroup generated by A, then $R(\lambda; A) x$ is the Laplace transform of $T(.)x$, and we conclude that $T(.) = S(.)$ by the Uniqueness Theorem for Laplace transforms.||||

PROOF OF THEOREM 1.38. For $0 \leq t \leq \tau$, we have by Lemma 3. $||S_n(t)|| \leq M e^{r\tau} q^n$, with $q < 1$. Therefore the series

$$S(t) := \Sigma_{n \geq 0} S_n(t) \qquad (**)$$

converges in $B(X)$-norm, uniformly on every interval $[0, \tau]$. By Lemma 3., it follows that $S(.)$ is strongly continuous on $[0, \infty)$, and $||S(t)|| \leq \frac{M}{1-q} e^{rt}$.

Since $S_n(0) = 0$ for $n \geq 1$ (cf. Lemma 3.), we have $S(0) = I$.

The exponential growth of $||S(.)||$ shows that the Laplace transform of $S(.)x$ converges absolutely for $\Re \lambda > r$, and a routine application of the Lebesgue Dominated Convergence Theorem shows that

$$\int_0^\infty e^{-\lambda t} S(t) x \, dt = \Sigma_{n \geq 0} \int_0^\infty e^{-\lambda t} S_n(t) x \, dt$$

$$= \Sigma_n R(\lambda; A) [B R(\lambda; A)]^n x = R(\lambda; A + B) x$$

by Lemma 4. and Lemma 2., and we conclude now from Lemma 5. that $A + B$ generates the C_o-semigroup $S(.)$ (given explicitly by $(**)$ and Lemma 3.).||||

G. GROUPS AND SEMI-SIMPLICITY MANIFOLD

If the C_o-semigroup $T(.)$ can be extended to $\mathbb{R} = (-\infty, \infty)$ with preservation of the identity

$$T(s)T(t) = T(s+t) \qquad (s, t \in \mathbb{R}),$$

it will be called a C_o-group of operators.

When this is the case, the semigroup $S(t) := T(-t)$, $t \geq 0$, is also of class C_o, since for $0 < t < \delta$,

$$S(t)x - x = T(-\delta)[T(\delta - t)x - T(\delta)x] \to_{t \to 0+} 0$$

for all $x \in X$ (cf. Theorem 1.1).

The generator A' of $S(.)$ is $-A$, because for $x \in D(A)$,

$$t^{-1}[S(t)x - x] = -T(-\delta)(-t)^{-1}[T(\delta - t)x - T(\delta)x] \to_{t \to 0+} -Ax$$

by Theorem 1.2, so that $-A \subset A'$, and therefore $A' = -A$ by symmetry.

Let ω, ω' be the types of $T(.)$ and $S(.)$ respectively. Since $||S(t)|| = ||T(t)^{-1}|| \geq ||T(t)||^{-1}$, we have

$$\omega' = \lim_{t \to \infty} t^{-1} \log ||S(t)|| \geq - \lim_{t \to \infty} t^{-1} \log ||T(t)|| = -\omega.$$

Note also that

$$\omega' = - \lim_{t \to -\infty} t^{-1} \log ||T(t)||.$$

By Theorem 1.15, since $\sigma(-A) = -\sigma(A)$, the spectrum of A is necessarily contained in the closed strip

$$S : -\omega' \leq \Re\lambda \leq \omega. \tag{*}$$

Also since $R(\lambda; -A) = -R(-\lambda; A)$, the necessary condition

$$||R(\lambda; -A)^n|| \leq M/(Re\lambda - \omega')^n$$

for $\Re\lambda > \omega'$ becomes $||R(\lambda; A)^n|| \leq M/[(-\omega') - Re\lambda]^n$ for $\Re\lambda < -\omega'$. Thus the growth condition on the resolvent outside the strip (*) is

$$||R(\lambda; A)^n|| \leq M/d_\lambda^n, \tag{**}$$

40

where $d_\lambda = d(\lambda, S)$ denotes the distance of λ to the strip S.

The sufficiency of (**) (together with the usual conditions that A be closed and densely defined) for A to generate a C_o-group is easily obtained from the Hille-Yosida Theorem. Indeed, applying the theorem separately in the half-planes $\Re\lambda > \omega$ and $\Re\lambda > \omega'$, we obtain two C_o-semigroups $T(.)$ and $S(.)$ with the respective generators A and $-A$. Using Theorem 1.2, we have for all $x \in D(A)$:

$$[T(s)S(s)x]' = T(s)AS(s)x - T(s)S(s)Ax = 0 \qquad (s \geq 0).$$

Therefore $T(s)S(s) = T(0)S(0) = I$, and similarly $S(s)T(s) = I$, i.e., $S(s) = T(s)^{-1}$ for all $s \geq 0$. Thus A generates the C_o-group $T(.)$. Formally

1.39. THEOREM. The operator A generates a C_o-group of operators if and only if it is closed, densely defined, has spectrum in a strip $S : -\omega' \leq \Re\lambda \leq \omega$, and

$$||R(\lambda; A)^n|| \leq M/d(\lambda; S)^n$$

for all real $\lambda \notin [-\omega', \omega]$.

Consider now a *bounded* group:

$$||T(t)|| \leq M \qquad (t \in R).$$

In this case, both types vanish, so that

$$\sigma(A) \subset iR.$$

When X is a Hilbert space (with inner product $(.,.)$), the case of a bounded group reduces to that of a *unitary* group:

1.40. THEOREM. Let $T(.)$ be a bounded strongly continuous group of operators acting on the Hilbert space X. Then there exists a non-singular bounded (positive) operator Q such that $QT(.)Q^{-1}$ is a group of unitary operators.

PROOF. Let $\mathbb{B}(\mathbb{R})$ denote the Banach algebra of all bounded complex functions on $\mathbb{R}$, with the supremum norm, and let LIM be the "generalized Banach limit" functional on it (cf. [DS-I]). Define

$$(x,y)_T = LIM\,(T(t)x, T(t)y) \qquad (x,y \in X),$$

and let $||x||_T = (x,x)_T^{1/2}$. Then

$$M^{-1}||.|| \leq ||.||_T \leq M||.||,$$

so that X is a Hilbert space under the equivalent inner product $(.,.)_T$, and there exists therefore a (strictly) positive operator P such that $(x,y)_T = (Px, y)$ for all $x, y \in X$. Let Q be the positive square root of P. For $s \in \mathbb{R}$ and $x, y \in X$ fixed, write $x = Qu$ and $y = Qv$. Then

$$(QT(s)Q^{-1}x, QT(s)Q^{-1}y) = (PT(s)u, T(s)v) = (T(s)u, T(s)v)_T$$

$$= LIM_t(T(t)T(s)u, T(t)T(s)v) = LIM_t(T(t+s)u, T(t+s)v) = LIM\,(T(t)u, T(t)v)$$

$$= (u,v)_T = (Pu, v) = (Qu, Qv) = (x,y).\|\|\|$$

We make a short elementary digression about symmetric operators on a Hilbert space X. If A is a densely defined operator on X, we set $D(A^*)$ to be the set of all $y \in X$ for which there exists a (necessarily unique) $z \in X$ such that $(Ax, y) = (x, z)$ for all $x \in D(A)$; we then set $A^*y = z$ for all $y \in D(A^*)$. Thus

$$(Ax, y) = (x, A^*y) \qquad (x \in D(A), y \in D(A^*)).$$

The **Hilbert adjoint** A^* of A is closed [indeed, if $y_n \in D(A^*)$ converge to y and $A^*y_n \to z$, then

$$(Ax, y) = \lim(Ax, y_n) = \lim(x, A^*y_n) = (x, z),$$

so that $y \in D(A^*)$ and $A^*y = z$].

A densely defined operator A is **symmetric** if

$$(Ax, y) = (x, Ay) \qquad (x, y \in D(A)).$$

This is of course equivalent to $A \subset A^*$. In particular, any symmetric operator is closable, and its closure A^- (which satisfies necessarily $A^- \subset A^*$) is a closed symmetric operator. We say that A is **selfadjoint** if $A = A^*$, and **essentially selfadjoint** if $A^- = A^*$. Note that we always have $A^* = (A^-)^*$ and $A^- = A^{**}$, so that A is essentially selfadjoint if and only if $A^* = A^{**}$.

As usual, $[D(A^*)]$ denotes the Hilbert space $D(A^*)$ with the (Hilbert) graph-norm, induced by the graph-inner- product

$$(x,y)_{A^*} := (x, y) + (A^*x, A^*y) \qquad (x, y \in D(A^*)).$$

Since $A^* : [D(A^*)] \to X$ is continuous, the subspaces

$$D_+ := ker\,(iI - A^*); \quad D_- := ker\,(-iI - A^*)$$

are closed subspaces of $[D(A^*)]$; $Ax = ix$ on D_+, $Ay = -iy$ on D_-, and the subspaces are clearly orthogonal:

$$(x,y)_{A^*} = (x,y) + (ix,-iy) = 0$$

for all $x \in D_+$ and $y \in D_-$. Again, for x,y as above and $z \in D(A^-)$ for A *symmetric* (so that $A^- \subset A^*$), we have

$$(z,x)_{A^*} = (z,x) + (A^*z, A^*x) = (z,x) + (A^-z, ix)$$

$$= (z,x) + (z, A^*ix) = (z,x) - (z,x) = 0,$$

and similarly for y. Therefore the (Hilbert) direct sum of D_+ and D_- is contained in the orthocomplement Y of $D(A^-)$ in $[D(A^*)]$. On the other hand, if $u \in Y$, then for all $y \in D(A)$,

$$0 = (y,u)_{A^*} = (y,u) + (Ay, A^*u),$$

so that $(Ay, A^*u) = (y,-u)$; hence $A^*u \in D(A^*)$, and $A^*(A^*u) = -u$. Therefore $(iI - A^*)(-iI - A^*)u = 0$ (with commuting factors!), showing that

$$(-iI - A^*)u \in D_+; \quad (iI - A^*)u \in D_-$$

for all $u \in Y$. Since $u = (1/2i)[(iI - A^*)u - (-iI - A^*)u]$, we see that Y is contained in the direct sum of D_+ and D_-, hence equals it (by the preceding observation). Thus

$$[D(A^*)] = D(A^-) \oplus D_+ \oplus D_-. \tag{1}$$

The subspaces D_+ and D_- are called the **deficiency subspaces** of A, and their Hilbert dimensions are the **deficiency indices** n_+ and n_- respectively. Since it is generally true, for any densely defined operator T, that $\ker T^*$ equals the orthocomplement of $\operatorname{ran} T$, we have

$$D_+ = [\operatorname{ran}(-iI - A)]^\perp; \quad D_- = [\operatorname{ran}(iI - A)]^\perp,$$

where the orthocomplement is taken in the space X. In particular, we read from the decomposition (1) that the symmetric operator A is essentially selfadjoint if and only if both $iI - A$ and $-iI - A$ have dense range in X.

After this digression about symmetric operators, consider again a C_o-group of unitary operators $T(.)$, and write its generator as $A = iH$. Then for all $x, y \in D(H) = D(A)$,

$$(Hx,y) = (-i)\lim_{t\to 0+} (t^{-1}[T(t) - I]x, y) = (-i)\lim(x, t^{-1}[T(-t) - I]y)$$

$$= -i(x, -Ay) = (x, Hy),$$

i.e., H is a (densely defined closed) symmetric operator. Also $\sigma(H) = -i\sigma(A) \subset \mathbb{R}$ (see above!), so that, in particular, $iI - H$ and $-iI - H$ are *onto*. By the preceding remarks, this proves that H is selfadjoint. Let e^{itH} be the operator associated with H by means of the operational calculus for selfadjoint operators:

$$e^{itH}x := \int_{\mathbb{R}} e^{its} E(ds)x, \qquad (2)$$

where $E(.)$ is the resolution of the identity for H. A trivial application of dominated convergence shows that e^{itH} is a C_o-group with generator $iH = A$. Therefore $T(t) = e^{itH}$. Since any C_o-semigroup of *unitary* operators extends in an obvious manner to a C_o-group, we have

1.41. THEOREM (Stone). Let $T(.)$ be a C_o-(semi)group of unitary operators in Hilbert space. Then there exists a selfadjoint operator H such that $T(t) = e^{itH}$ for all $t \in \mathbb{R}$.

Combining this with Theorem 1.40, we obtain

1.42. COROLLARY. Let $T(.)$ be a bounded C_o-group of operators in Hilbert space. Then there exists a (unique, not necessarily selfadjoint) spectral measure $E(.)$ on the Borel algebra of $\mathbb{R}$ such that

$$T(t) = \int_{\mathbb{R}} e^{its} E(ds) \qquad (t \in \mathbb{R}),$$

where the integral exists in the strong operator topology.

Using the terminology of [DS-III], the generator of $T(.)$ equals iS, where S is a scalar-type spectral operator with real spectrum, and $T(t) = e^{itS}$ (using the operational calculus for S). Also QSQ^{-1} is selfadjoint, with Q as in Theorem 1.40.

REMARK. A neat way to deal with the Hilbert adjoint is to consider its graph $G(A^*) \subset X^2$, where the cartesian product X^2 is a Hilbert space under the inner product

$$([x, y], [x', y']) := (x, x') + (y, y').$$

The operator

$$J : [x, y] \to [y, -x]$$

is a unitary operator on X^2, and a simple calculation shows that

$$G(A^*) = \{JG(A)\}^{\perp}.$$

One reads from this formula that A^* is closed (its graph is closed as an orthocomplement!), and that $A^* = (A^-)^*$. Also, since J is unitary with $J^2 = I$,

$$G(A^{**}) = [JG(A^*)]^\perp = JG(A^*)^\perp$$

$$= J[JG(A)]^{\perp\perp} = J[JG(A)]^- = J^2G(A)^- = G(A)^- = G(A^-).$$

Therefore $A^{**} = A^-$.

Stone's theorem can be used to prove Bochner's theorem about Fourier-Stieltjes' transforms.

1.43. THEOREM (Bochner). A continuous function $f : \mathbb{R} \to \mathbb{C}$ is the Fourier-Stieltjes transform of a finite positive Borel measure μ (i.e., $f(t) = \int_\mathbb{R} e^{its}\mu(ds)$) if and only if it is positive definite, that is,

$$\Sigma_{j,k} f(t_j - t_k)c_j\overline{c_k} \geq 0$$

for all *finite* sequences $t_j \in \mathbb{R}$ and $c_j \in \mathbb{C}$.

PROOF. Let X_o be the complex vector space of all complex functions on $\mathbb{R}$ with *finitely* many non-zero values. Let

$$(\phi, \psi) = \Sigma_{t,s\in\mathbb{R}} f(t - s)\phi(t)\psi(s)^-,$$

where f is a given positive definite function, and $\phi, \psi \in X_o$. This is a *pseudo-inner* product on X_o (i.e., we may have $(\phi, \phi) = 0$ for non-zero ϕ). The set $K = \{\phi \in X_o; (\phi, \phi) = 0\}$ is a closed subspace of X_o. The factor space $X_1 = X_o/K$ is a pre-Hilbert space with the inner product $(\phi + K, \psi + K) = (\phi, \psi)$ (which is well- defined, i.e., independent of the choice of the cosets representatives). Let X be the completion of the pre-Hilbert space X_1. Define U_r^o on X_o by $[U_r^o\phi](t) = \phi(t - r)$, $r \in \mathbb{R}$. Then $(U_r^o\phi, U_r^o\psi) = (\phi, \psi)$. Since U_r^o maps K into itself, it induces an operator U_r^1 of X_1 into itself, well-defined by $U_r^1(\phi + K) = U_r^o\phi + K$, and U_r^1 is a unitary operator of X_1 onto itself. It extends uniquely to a unitary operator U_r of X onto itself. The family $\{U_r; r \in \mathbb{R}\}$ is a C_o-group on X (where the C_o-property follows from the assumed *continuity* of f). By Stone's theorem, $(U_r x, x) = \int_\mathbb{R} e^{its}(E(ds)x, x)$ for all $x \in X$, where $(E(.)x, x) = ||E(.)x||^2$ is a finite positive Borel measure. In particular, for $x_o = \phi_o + K$ with $\phi_o(t) = 1$ for $t = 0$ and vanishing otherwise, we have

$$(U_r x_o, x_o) = (U_r^1 x_o, x_o) = (U_r^o\phi_o, \phi_o) = f(r),$$

so that f is indeed the Fourier-Stieltjes transform of the finite positive Borel measure $(E(.)x_o, x_o)$.

The converse is an easy calculation.||||

It is also possible to deduce Stone's theorem from Bochner's. If $T(.)$ is a C_o-group of unitary operators on the Hilbert space X, then $f := (T(.)x, x)$ is continuous and positive definite (for each $x \in X$):

$$\Sigma_{j,k} f(t_j - t_k) c_j c_k^- = \Sigma(T(t_k)^* T(t_j)x, x) c_j c_k^-$$

$$= \Sigma(T(t_j)x, T(t_k)x) c_j c_k^- = ||\Sigma_j c_j T(t_j)x||^2 \geq 0.$$

Therefore, by Bochner's theorem, there exists a family $\{\mu(.;x); x \in X\}$ of finite positive Borel measures on $\mathbb{R}$ such that

$$(T(t)x, x) = \int_{\mathbb{R}} e^{its} \mu(ds; x)$$

for all $t \in \mathbb{R}$ and $x \in X$.
   Define

$$\mu(.; x, y) := (1/4)\Sigma_{0 \leq k \leq 3} i^k \mu(.; x + i^k y) \qquad (x, y \in X).$$

These are (finite) complex Borel measures, and

$$(T(t)x, y) = \int_{\mathbb{R}} e^{its} \mu(ds; x, y) \qquad (t \in \mathbb{R}; x, y \in X).$$

From this representation and the uniqueness property of the Fourier-Stieltjes transform, it is easy to deduce the existence of a resolution of the identity E such that $T(t) = \int e^{its} E(ds) = e^{itA}$ for the selfadjoint operator $A := \int sE(ds)$ with domain $D(A) = \{x \in X; \int s^2(E(ds)x, x) < \infty\}$. The details are given in an analogous proof below, with X a reflexive Banach space.
   For the Banach space setting, we start with another theorem of Bochner, characterizing the Fourier-Stieltjes transforms of complex Borel measures.

1.44. THEOREM (Bochner). A function $f : \mathbb{R} \to \mathbb{C}$ is the Fourier-Stieltjes transform of a complex regular Borel measure μ with $||\mu|| \leq K$ if and only if it is continuous and

$$|\Sigma_j c_j f(t_j)| \leq K||\Sigma_j c_j e^{it_j s}||_\infty$$

for all finite sequences of real t_j and complex c_j.

PROOF. See [R2], p. 32.

We consider the normed space $\mathbb{P} = \mathbb{P}(\mathbb{R})$ of all "trigonometric polynomials"

$$\phi(s) = \Sigma_j c_j e^{it_j s} \qquad (c_j \in \mathbb{C}; t_j \in \mathbb{R}),$$

where the sum above is *finite*, with the supremum norm. Let Y be any Banach space. Given a function $f : \mathbb{R} \to Y$, we define the linear operator $B_f : \mathbb{P} \to Y$ by

$$B_f \phi := \Sigma_j c_j f(t_j),$$

for $\phi \in \mathbb{P}$ as above, and set

$$||f||_B := ||B_f||,$$

where the norm on the right is the operator norm (that could be infinite a priori). We refer to $||.||_B$ as the "Bochner norm", and consider the vector space

$$\mathbb{F}(Y) := \{f : \mathbb{R} \to Y; ||f||_B < \infty\}.$$

Let $\phi_t(s) = e^{its}$, $\quad t, s \in \mathbb{R}$. Then for any $f : \mathbb{R} \to Y$,

$$||f||_B = ||B_f|| \geq ||B_f \phi_t|| = ||f(t)|| \qquad (t \in \mathbb{R}),$$

so that

$$||f||_B \geq ||f||_\infty := \sup_t ||f(t)||. \qquad (1)$$

Thus all functions in $\mathbb{F}(Y)$ are bounded, and $||.||_B$ is a norm on that space. By (1), if $\{f_n\} \subset \mathbb{F}(Y)$ is $||.||_B$-Cauchy, it converges uniformly in Y to some f. Given $\epsilon > 0$, let n_o be such that $||f_n - f_m||_B < \epsilon$ for all $n > n_o$. Then

$$||\Sigma_j c_j [f_n(t_j) - f_m(t_j)]|| \leq \epsilon$$

for all c_j, t_j such that the corresponding ϕ has supremum norm equal to 1, and all $n, m > n_o$. Letting $m \to \infty$, we get

$$||\Sigma_j c_j [f_n(t_j) - f(t_j)]|| \leq \epsilon$$

for all $n > n_o$, and all c_j, t_j as above. Thus $||f_n - f||_B \leq \epsilon < \infty$ (for $n > n_o$), i.e., $f_n - f \in \mathbb{F}(Y)$, and so $f = f_n - (f_n - f) \in \mathbb{F}(Y)$, and $f_n \to f$ in the Bochner norm. This shows that $\mathbb{F}(Y)$ is a Banach space with the Bochner norm.

A simple calculation shows that $\mathbb{F}(Y)$ contains all Fourier-Stieltjes transforms of Y-valued vector measures m (cf. [DS-I], Section IV.10): if $f(t) := \int_{\mathbb{R}} e^{its} m(ds)$, then for all ϕ as above with $||\phi||_\infty = 1$,

$$||\Sigma_j c_j f(t_j)|| = ||\int_{\mathbb{R}} \phi(s) m(ds)|| \leq ||m||,$$

so that $||f||_B \leq ||m|| < \infty$, where $||m||$ denotes the "semi-variation" of m.

In particular, $\mathbb{F}(Y)$ contains the constant functions c, and $||c||_B = ||c||$.

The Bochner norm is invariant under additive translation $f(t) \to f(t+c)$ and non-zero multiplicative translation $f(t) \to f(ct)$ $(c \in \mathbb{R})$.

If Y, Z are Banach spaces and $U \in B(Y, Z)$, then $U\mathbb{F}(Y) \subset \mathbb{F}(Z)$ and

$$||Uf||_B \le ||U|| \cdot ||f||_B \qquad (f \in \mathbb{F}(Y)).$$

We have

$$||f||_B = \sup\{||y^* f||_B ; y^* \in Y^*, ||y^*|| = 1\}.$$

By the Uniform Boundedness Theorem, $f \in \mathbb{F}(Y)$ if and only if $y^* f \in \mathbb{F}(\mathbb{C})$ for all $y^* \in Y^*$.

Also, if $F \to B(X, Y)$ for Banach spaces X, Y, then the following are equivalent:
(i) $F \in \mathbb{F}(B(X, Y))$;
(ii) $Fx \in \mathbb{F}(Y)$ for all $x \in X$;
(iii) $y^* Fx \in \mathbb{F}(\mathbb{C})$ for all $x \in X, y^* \in Y^*$,
In addition,

$$||F||_B = \sup\{||Fx||_B ; x \in X, ||x|| = 1\}$$

$$= \sup\{||y^* Fx||_B ; x \in X, y^* \in Y^*, ||x|| = ||y^*|| = 1\}.$$

If $f \in \mathbb{F}(Y)$ is weakly *continuous*, then by Bochner's theorem, there corresponds to each $y^* \in Y^*$ a finite regular complex Borel measure $\mu(.; y^*)$ such that

$$y^* f(t) = \int_{\mathbb{R}} e^{its} \mu(ds; y^*) \qquad (t \in \mathbb{R}),$$

and

$$||\mu(.; y^*)|| \le ||f||_B ||y^*||.$$

The uniqueness of the integral representation implies that for each $\delta \in \mathcal{B}(\mathbb{R})$ (the Borel algebra of $\mathbb{R}$), $\mu(\delta; .)$ is a linear functional on Y^* with norm $\le ||f||_B$. Therefore

$$\mu(\delta; y^*) = m(\delta) y^* \qquad (y^* \in Y^*, \delta \in \mathcal{B}(\mathbb{R})),$$

where $m : \mathcal{B}(\mathbb{R}) \to Y^{**}$ is such that $m(.) y^*$ is a regular finite Borel measure for each $y^* \in Y^*$. The element $\int_{\mathbb{R}} e^{its} m(ds) \in Y^{**}$, well-defined by the relation

$$[\int_{\mathbb{R}} e^{its} m(ds)] y^* = \int_{\mathbb{R}} e^{its} [m(ds) y^*] \qquad (y^* \in Y^*),$$

coincides with the element $f(t) \in X$ (imbedded in Y^{**} as usual). In this weakened sense, the weakly continuous elements of $\mathbb{F}(Y)$ are Fourier-Stieltjes transforms of Y^{**}- valued measures like m. When Y is reflexive, m is a weakly countably additive (hence strongly countably additive, by Pettis' theorem, cf. [DS-I]) Y-valued measure, and we can write

$$f(t) = \int_{\mathbb{R}} e^{its} m(ds) \qquad (t \in \mathbb{R}),$$

so that the weakly continuous functions in $\mathbb{F}(Y)$ are precisely the Fourier-Stieltjes transforms of vector measures m as above. We summarize the above discussion formally:

1.45. PROPOSITION. The space $\mathbb{F}(Y)$ is a Banach space for the Bochner norm, and contains all the Fourier-Stieltjes transforms of Y-valued measures with finite variation. Every weakly continuous function in the space is the Fourier-Stieltjes transform of a Y^{**}-valued measure m such that $m(.)y^*$ is a regular complex Borel measure for each $y^* \in Y^*$. When Y is reflexive, every weakly continuous function in the space is the Fourier-Stieltjes transform of a strongly countably additive Y-valued measure m (such that $y^*m(.)$ is a regular complex Borel measure for each y^*), and is in particular strongly continuous as well.

1.46. DEFINITION. Suppose iA generates a C_o-group $T(.)$ on the Banach space X. We set

$$||x||_T = ||T(.)x||_B \qquad (x \in X).$$

The "semi-simplicity manifold" for $T(.)$ is the set

$$Z = Z_T = \{x \in X; ||x||_T < \infty\}.$$

1.47. LEMMA. $(Z, ||.||_T)$ is a Banach subspace of X, invariant for any $U \in B(X)$ which commutes with $T(.)$. Also $Z = X$ if and only if $||T(.)||_B < \infty$ (in this case, the two norms on X are equivalent).

PROOF. Since $||x||_T \geq ||T(.)x||_\infty \geq ||x||$, $(Z, ||.||_T)$ is a normed space. We prove completeness. If $\{x_n\}$ is Cauchy in $(Z, ||.||_T)$, it is also Cauchy in X; let x be its X-limit. Then $T(t)x_n \to T(t)x$ for each $t \in \mathbb{R}$. By definition of the $||.||_T$-norm, $\{T(.)x_n\}$ is Cauchy in the Banach space $(\mathbb{F}(X), ||.||_B)$. Therefore $T(.)x_n \to f$ in that space, and since $||.||_B \geq ||.||_\infty$, $f(t) = \lim_n T(t)x_n = T(t)x$ (limit in X), for each t. Thus $T(.)x \in \mathbb{F}(X)$, i.e., $x \in Z$, and $||x_n - x||_T = ||T(.)x_n - T(.)x||_B \to 0$, and Z (with the $||.||_T$-norm) is a Banach subspace of X.

If $U \in B(X)$ commutes with $T(t)$ for each $t \in \mathbb{R}$, then for each $x \in Z$, we have $T(.)x \in \mathbb{F}(X)$, and therefore $UT(.)x \in \mathbb{F}(X)$ and $||UT(.)x||_B \leq ||U||.||T(.)x||_B$ (see above), which is equivalent in our present situation to $T(.)[Ux] \in \mathbb{F}(X)$ (i.e., $Ux \in Z$) and $||Ux||_T \leq ||U||.||x||_T$. This shows that Z is U-invariant, and $||U||_{B(Z)} \leq ||U||_{B(X)}$. If $||T(.)||_B < \infty$, then for all trigonometric polynomials ϕ as above, and for all $x \in X$, $||\Sigma_j c_j T(t_j)x|| \leq ||\Sigma_j c_j T(t_j)||.||x|| \leq ||T(.)||_B||x||$, and therefore $||x||_T \leq ||T(.)||_B||x|| < \infty$, i.e., $Z = X$. Conversely, if $Z = X$, then $\sup ||[\Sigma_j c_j T(t_j)]x|| < \infty$ (supremum over all ϕ as above), and therefore $||T(.)||_B := \sup ||\Sigma_j c_j T(t_j)|| < \infty$ by the Uniform Boundedness Theorem. The equivalence of the norms is a consequence of the Closed Graph Theorem, or explicitly from the above discussion,

$$||x|| \le ||x||_T \le ||T(.)||_B ||x|| \qquad (x \in X).||||$$

1.48. DEFINITION. Let W be a linear manifold in X, and let $\mathcal{T}(W)$ denote the algebra of all operators with domain W and range in W.

A "spectral measure on W" is a function

$$E(.): \mathcal{B}(\mathbb{R}) \to \mathcal{T}(W),$$

such that

(i) $E(\mathbb{R}) = I/W$;

(ii) for each $x \in W$, $E(.)x$ is a regular, strongly countably additive vector measure; and

(iii) $E(\delta \cap \epsilon) = E(\delta)E(\epsilon)$, for all $\delta, \epsilon \in \mathcal{B}(\mathbb{R})$.

Note that by [DS-I] (Corollary III.4.5), $E(.)x$ is necessarily bounded for each $x \in W$.

Note also that (ii) is equivalent to

(ii') for each $x \in W$ and $x^* \in X^*$, $x^* E(.)x$ is a regular complex Borel measure. This follows from Pettis' theorem (cf. [DS-I]).

Let $\mathbb{B}(\mathbb{R})$ denote the Banach algebra of all bounded complex Borel functions on $\mathbb{R}$. For E as above and $h \in \mathbb{B}(\mathbb{R})$, the operator $\tau(h): W \to X$ is defined by

$$\tau(h)x = \int_{\mathbb{R}} h(s)E(ds)x \qquad (x \in W).$$

We then extend τ to the algebra $\mathbb{B}_{loc}(\mathbb{R})$ of all complex Borel functions on $\mathbb{R}$ that are bounded on each interval $[a, b]$, by letting

$$\tau(h)x = \lim_{a,b} \int_a^b h(s)E(ds)x := \int_{\mathbb{R}} h(s)E(ds)x$$

for $h \in \mathbb{B}_{loc}(\mathbb{R})$, where $\lim_{a,b}$ stands for the limit in X when $a \to -\infty$ and $b \to \infty$; the domain of $\tau(h)$ is the set of all $x \in W$ for which the limit exists.

We are now ready to state our Banach space version of the Stone Theorem.

1.49. THEOREM. Let $T(.)$ be a C_o-group of operators on the *reflexive* Banach space X, with generator iA and $\sigma(A) \subset \mathbb{R}$. Let Z be the semi-simplicity manifold for $T(.)$. Then there exists a spectral measure E on Z with the following properties:

(1) $T(t)x = \int_{\mathbb{R}} e^{its}E(ds)x \qquad (x \in Z; t \in \mathbb{R})$;

(2) E commutes with every $U \in B(X)$ commuting with $T(.)$;

(3) τ (corresponding to E) is a norm-decreasing algebra homomorphism of $\mathbb{B}(\mathbb{R})$ into $B(Z, ||.||_T)$, such that $\tau(\phi_t) = T(t)/Z$ for $\phi_t(s) = e^{its}$, $s, t \in \mathbb{R}$;

(4) If $f_1(s) = s$ $(s \in \mathbb{R})$, then $A_Z = \tau(f_1)_Z$ (where the subscript Z means the *part* of the relevant operator in Z), that is,

(i) $D(A_Z) = \{x \in Z; \int_{\mathbb{R}} sE(ds)x$ exists and belongs to $Z\}$, and

(ii) $Ax = \int_{\mathbb{R}} sE(ds)x$ $(x \in D(A_Z))$.

In addition, Z is maximal and E is unique in the following sense: if W is a Banach subspace of X and F is a spectral measure on W for which (3) is valid, then W is a Banach subspace of Z and $F(.) = E(.)/W$.

PROOF. For each $x \in Z$, the function $T(.)x$ is a strongly continuous element of $\mathbb{F}(X)$. Since X is reflexive, Proposition 1.45 gives a strongly countably additive X-valued measure $m(.; x)$ on $\mathcal{B}(\mathbb{R})$ such that

$$T(t)x = \int_{\mathbb{R}} e^{its} m(ds; x) \tag{1}$$

for all $t \in \mathbb{R}, x \in Z$;

$$||m(.; x)|| \leq ||x||_T; \tag{2}$$

and $x^* m(.; x)$ is a regular complex Borel measure for each $x^* \in X^*$. The uniqueness property of the Fourier-Stieltjes transform of regular complex Borel measures and the linearity of the left side of (1) imply that $m(.; x) = E(.)x$, where $E(\delta)$ is a linear transformation from Z to X, for each $\delta \in \mathcal{B}(\mathbb{R})$. By (1) with $t = 0$, $E(\mathbb{R}) = I/Z$. We rewrite (1) in the form

$$T(t)x = \int_{\mathbb{R}} e^{its} E(ds)x \qquad (t \in \mathbb{R}, x \in Z). \tag{1'}$$

If $U \in B(X)$ commutes with $T(.)$ and $x \in Z$, then $Ux \in Z$ by Lemma 1.47, and by (1'),

$$\int e^{its} U E(ds)x = UT(t)x = T(t)Ux = \int e^{its} E(ds)Ux,$$

hence $UE(\delta)x = E(\delta)Ux$ for all $x \in Z, \delta \in \mathcal{B}(\mathbb{R})$.

For each $\phi, \psi \in \mathbb{P}$ with $||\phi||_\infty = ||\psi||_\infty = 1$ and with respective parameters $c_j, t_j; c'_k, t'_k$, we have for $x \in Z$,

$$||\Sigma_k c'_k T(t'_k).\Sigma_j c_j T(t_j)x|| = ||\Sigma_{k,j} c'_k c_j T(t'_k + t_j)x||$$

$$\leq ||x||_T ||\Sigma_{k,j} c'_k c_j e^{is(t'_k + t_j)}||_\infty \leq ||x||_T ||\phi||_\infty ||\psi||_\infty = ||x||_T.$$

Therefore

$$||\Sigma_j c_j T(t_j)x||_T \leq ||x||_T$$

for all parameters as above. Fix $h \in \mathbb{B}(\mathbb{R}), x \in Z$. Since $\Sigma_j c_j T(t_j)$ is a bounded operator commuting with $T(.)$, we have by (2)

$$||\Sigma_j c_j T(t_j)\tau(h)x|| = ||\tau(h)\Sigma_j c_j T(t_j)x||$$

$$\leq ||h||_\infty ||E(.)\Sigma_j c_j T(t_j)x|| \leq ||h||_\infty ||\Sigma_j c_j T(t_j)x||_T \leq ||h||_\infty ||x||_T.$$

Therefore

$$||\tau(h)x||_T \leq ||h||_\infty ||x||_T$$

for all $h \in \mathbb{B}(\mathbb{R}), x \in Z$. In particular

$$\tau : \mathbb{B}(\mathbb{R}) \to B(Z, ||.||_T)$$

has norm ≤ 1 ; actually, $||\tau|| = 1$, because

$$||\tau(\phi_r)x||_T = ||T(r)x||_T := ||T(.)T(r)x||_B$$

$$= ||T(. + r)x||_B = ||T(.)x||_B = ||x||_T,$$

by the translation invariance of the Bochner norm on $\mathbb{F}(\mathbb{R})$.

Taking $h = \chi_\delta$ (the characteristic function of $\delta \in \mathcal{B}(\mathbb{R})$), we get that

$$||E(\delta)||_{B(Z)} \leq 1,$$

so that surely $E(\delta) \in \mathcal{T}(Z)$. For $t, u \in \mathbb{R}$ and $x \in Z$, with all integrals below extending over $\mathbb{R}$, we have

$$\int e^{ius} E(ds)T(t)x = T(u)T(t)x = T(u+t)x$$

$$= \int e^{ius}[e^{its} E(ds)x].$$

By uniqueness for Fourier-Stieltjes transforms,

$$E(\delta)T(t)x = \int e^{its} \chi_\delta(s) E(ds)x$$

for all $\delta \in \mathcal{B}(\mathbb{R})$, etc. However the left side equals $T(t)E(\delta)x = \int e^{its} E(ds)E(\delta)x$ since $E(\delta)x \in Z$ for $x \in Z$. Therefore, again by uniqueness for Fourier-Stieltjes transforms,

$$E(ds)E(\delta)x = \chi_\delta E(ds)x, \tag{3}$$

so that

$$E(\sigma)E(\delta)x = \int \chi_\sigma \chi_\delta E(ds)x = E(\sigma \cap \delta)x$$

52

for all $\sigma, \delta \in \mathcal{B}(\mathbb{R})$ and $x \in Z$.

We conclude that E is a spectral measure on Z.

By (3),

$$\tau(h)\tau(\chi_\delta)x = \tau(h)E(\delta)x = \int h(s)\chi_\delta(s)E(ds)x = \tau(h\chi_\delta)x$$

for all $h \in \mathbb{B}(\mathbb{R}), \delta \in \mathcal{B}(\mathbb{R}), x \in Z$. By linearity of τ, it follows that $\tau(hg) = \tau(h)\tau(g)$ for all $h \in \mathbb{B}(\mathbb{R})$ and $g \in \mathbb{B}_o(\mathbb{R})$, the subalgebra of simple Borel functions. Next, for $g \in \mathbb{B}(\mathbb{R})$, choose simple Borel functions g_n converging uniformly to g. Then for all $x \in Z$,

$$||\tau(hg)x - \tau(h)\tau(g)x|| \le ||\tau[h(g - g_n)]x|| + ||\tau(h)\tau(g_n - g)x||$$

$$\le ||h(g - g_n)||_\infty ||x||_T + ||h||_\infty ||\tau(g_n - g)x||_T$$

$$\le 2||h||_\infty ||g_n - g||_\infty ||x||_T \to 0$$

as $n \to \infty$, and Statement (3) of the theorem is proved.

For all $t \in \mathbb{R}$, $\sigma(itA) = it\sigma(A) \subset i\mathbb{R}$, so that $R(t) := R(1; itA)$ is a well-defined bounded operator commuting with $T(.)$. If $x \in R(t)Z$, say $x = R(t)z$ with $z \in Z$, then $x \in D(A) \cap Z$, and $Ax = (it)^{-1}(x - z) \in Z$ (for $t \ne 0$), i.e., $R(t)Z \subset D(A_Z)$. On the other hand, if $x \in D(A_Z)$, then $z := (1 - itA)x \in Z$, and therefore $x = R(t)z \in R(t)Z$. This shows that

$$D(A_Z) = R(t)Z \qquad (0 \ne t \in \mathbb{R}). \tag{4}$$

Let $x \in D(A_Z)$; write then $x = R(t)z$ with $z \in Z$ and $0 \ne t \in \mathbb{R}$ fixed. By Theorem 1.15,

$$R(t)z = t^{-1}R(t^{-1}; iA)z = t^{-1} \int_0^\infty e^{-s/t}T(s)z\,ds$$

$$= \int_0^\infty e^{-u}T(tu)z\,du \tag{5}$$

for all $0 \ne t \in \mathbb{R}, z \in X$. For $z \in Z$, we get

$$R(t)z = \int_0^\infty e^{-u} \int_\mathbb{R} e^{itus}E(ds)z\,du$$

$$= \int_\mathbb{R} \int_0^\infty e^{-u(1-its)}du\,E(ds)z = \int (1 - its)^{-1}E(ds)z, \tag{6}$$

where the interchange of the order of integration is justified by applying on both sides an arbitrary $x^* \in X^*$ and using Fubini's theorem.

For real $a < b$, we then have by (6) and the multiplicativity of τ on $\mathbb{B}(\mathbb{R})$ (for $x = R(t)z$ as above):

$$\int_a^b sE(ds)x = \int_a^b s(1 - its)^{-1}E(ds)z \to \int_{\mathbb{R}} s(1 - its)^{-1}E(ds)z$$

when $a \to -\infty$ and $b \to \infty$. Writing $s(1 - its)^{-1} = it^{-1}[1 - (1 - its)^{-1}]$, the last integral is seen to equal

$$it^{-1}[z - R(t)z] = it^{-1}(z - x) = Ax \in Z.$$

This shows that $D(A_Z) \subset \{x \in Z; \int_{\mathbb{R}} sE(ds)x$ exists and belongs to $Z\}$, and $Ax = \int sE(ds)x$ on $D(A_Z)$.

On the other hand, if x belongs to the set on the right of (i) (in the statement of the theorem), then denoting the integral in (i) by $z \in Z$, we obtain from the multiplicativity of τ for $t \neq 0$:

$$R(t)z = R(t)\lim_{a,b} \int_a^b sE(ds)x = \lim_{a,b} R(t) \int_a^b sE(ds)x$$

$$= \lim_{a,b} \int_a^b s(1 - its)^{-1}E(ds)x = \int_{\mathbb{R}} s(1 - its)^{-1}E(ds)x = it^{-1}[x - R(t)x]$$

(cf. preceding calculation).

Therefore $x = R(t)(x - itz) \in R(t)Z = D(A_Z)$ by (4), and we conclude that Property (4) in the statement of the theorem is valid.

Finally, suppose $(W, ||.||_W)$ is a Banach subspace of X for which Property (3) (in the statement of the theorem) is valid, with $(W, ||.||_W)$ replacing $(Z, ||.||_T)$ and $\tau' : \mathbb{B}(\mathbb{R}) \to B(W, ||.||_W)$ (induced by F) replacing τ. Then for all $\phi \in \mathbb{P}$ with $||\phi||_\infty = 1$ and parameters c_j, t_j,

$$||\Sigma_j c_j T(t_j)x|| = ||\tau'(\phi)x|| \leq ||\tau'(\phi)x||_W$$

$$\leq ||\tau'(\phi)||_{B(W)}||x||_W \leq ||x||_W \qquad (x \in W).$$

Therefore $||x||_T \leq ||x||_W$, and W is a Banach subspace of Z. Since $T(t)x = \int e^{its}F(ds)x = \int e^{its}E(ds)x$ for $x \in W$, the uniqueness property of the Fourier-Stieltjes transform implies that $F(.)x = E(.)x$ for all $x \in W$.||||

We consider the special case $Z = X$. By Lemma 1.47, this happens if and only if $||T(.)||_B < \infty$, and in this case the two norms $||.||$ and $||.||_T$ on X are equivalent. Let E be the spectral measure on $Z = X$ provided by Theorem 1.49. Since $||E(\delta)x||_T \leq ||x||_T$ for all $x \in X$, the equivalence of the norms shows that $E : \mathcal{B}(\mathbb{R}) \to B(X)$ is a "spectral measure" in the usual sense, that is, an algebra homomorphism of the Boolean algebra $\mathcal{B}(\mathbb{R})$ into $B(X)$ such that $E(.)x$ is regular and countably additive for each $x \in X$. Properties (4)(i),(ii) become

(i) $D(A) = \{x \in X; \int_{\mathbb{R}} sE(ds)x := \lim_{a,b} \int_a^b sE(ds)x \quad \text{exists}\}$; and

(ii) $Ax = \int_{\mathbb{R}} sE(ds)x \qquad (x \in D(A))$.

Using the terminology of [DS-III], the operator A is spectral of scalar type (with real spectrum). The map τ defined above is now the usual operational calculus for the scalar-type spectral operator A, and in particular, the semigroup $T(.)$ is precisely e^{itA}, as defined through this operational calculus. Note that when X is a Hilbert space, the condition $||T(.)||_\infty < \infty$ was necessary and sufficient for the above conclusions (Corollary 1.42), while our generalization to reflexive Banach space requires the stronger assumption $||T(.)||_B < \infty$. This latter condition is however necessary as well, by Proposition 1.45 and Lemma 1.47.

We formalize the above discussion in

1.50. COROLLARY. Let iA generate a C_o-group $T(.)$ on the reflexive Banach space X. Then A is a scalar-type spectral operator with real spectrum if and only if $||T(.)||_B < \infty$. In that case, $T(t) = e^{itA} := \int_{\mathbb{R}} e^{its} E(ds)$, where E is the resolution of the identity for A.

Actually, we can restate this corollary without assuming a priori that iA generates a C_o-group. We need only to assume that A is densely defined, and has real spectrum. Let then

$$R(t) := (I - itA)^{-1} \qquad (t \in \mathbb{R}).$$

We first establish some identities.

1.51. THEOREM. If iA (with $\sigma(A)$ real) generates a C_o-group $T(.)$, then

$$||x||_T = \sup_{n \geq 0} ||R^n x||_B \qquad (x \in X) \tag{*}$$

and

$$||T(.)||_B = \sup_{n \geq 0} ||R^n||_B.$$

PROOF. By Theorem 1.36,

$$T(t)x = \lim_{n \to \infty} R^n(t/n)x \qquad (x \in X; t \in \mathbb{R}).$$

Therefore for each $\phi \in \mathbb{P}$ with $||\phi||_\infty = 1$ and parameters c_j, t_j,

$$||\Sigma_j c_j T(t_j)x|| = \lim_n ||\Sigma_j c_j R^n(t_j/n)x||.$$

Each norm on the right is $\le ||R^n(./n)x||_B = ||R^n x||_B$, by the invariance of the Bochner norm under multiplicative translations, and this implies the inequality $\le$ in (*).

On the other hand, the Taylor expansion of the resolvent obtained in the proof of Theorem 1.11 shows that

$$R(\lambda; A)^{(n-1)} = (-1)^{n-1}(n-1)! R(\lambda; A)^n \qquad (\lambda \in \rho(A); n \ge 1). \tag{1}$$

The derivatives may be calculated by using Theorem 1.15, when A generates a C_o-semigroup $T(.)$. We then obtain the following Laplace transform representation for the powers of the resolvent:

$$R(\lambda; A)^n x = \int_0^\infty e^{-\lambda t}[t^{n-1}/(n-1)!]T(t)x \, dt, \tag{2}$$

for all $x \in X, \Re\lambda > \omega$, and $n \ge 1$.

In our case, with the generators iA and $-iA$ of the semigroups $T(.)$ and $S(t) := T(-t)$ $(t \ge 0)$ respectively, a simple calculation leads to the formula

$$R^n(t)x = \Gamma(n)^{-1} \int_0^\infty e^{-s} s^{n-1} T(ts)x \, ds, \tag{3}$$

for all $x \in X, n = 1, 2, \dots$ and $t \in \mathbb{R}$.

Since $||T(ts)x||_B = ||T(t)x||_B := ||x||_T$ for each fixed $s > 0$, we have for all ϕ as above,

$$||\Sigma_j c_j R^n(t_j)x|| = \Gamma(n)^{-1} ||\int_0^\infty e^{-s} s^{n-1} \Sigma_j c_j T(t_j s)x \, ds||$$

$$\le \Gamma(n)^{-1} \int_0^\infty e^{-s} s^{n-1} ||T(ts)x||_B \, ds = ||x||_T,$$

hence $||R^n x||_B \le ||x||_T$ for all $n \ge 0$, and (*) follows. The second identity is then an elementary consequence.||||

We can restate now Corollary 1.50 without assuming a priori that iA is a generator.

1.52. COROLLARY. Let A be a densely defined operator with real spectrum, acting in the reflexive Banach space X. Then A is a scalar-type spectral operator if and only if

$$V_A := \sup_{n \ge 0} ||R^n||_B < \infty.$$

(in that case, iA generates the group e^{itA}, which is the Fourier-Stieltjes transform of the resolution of the identity for A).

56

PROOF. If $V_A < \infty$, we surely have $||R^n||_\infty \le V_A < \infty$ for all n. Since

$$\lambda R(\lambda; iA) = R(1/\lambda) \qquad (0 \ne \lambda \in \mathbb{R}),$$

we have

$$||[\lambda R(\lambda; iA)]^n|| \le V_A \qquad (n = 1, 2, ...; 0 \ne \lambda \in \mathbb{R}).$$

Also iA is closed (since $\rho(iA)$ is non-empty) and densely defined (by hypothesis). The conditions of the Hille-Yosida theorem for groups (Theorem 1.39) are therefore satisfied by the operator iA, with $\omega' = \omega = 0$. If $T(.)$ denotes the group generated by iA, we have $||T(.)||_B = V_A < \infty$, and Corollary 1.50 applies to establish that A is a scalar-type spectral operator.

Conversely, if A is scalar-type spectral, let $E : \mathcal{B}(\mathbb{R}) \to B(X)$ be its resolution of the identity. Then iA generates the C_o-group $T(.) = e^{itA} := \int_{\mathbb{R}} e^{its} E(ds)$. In particular $||T(.)||_B < \infty$ by Proposition 1.45, that is, $V_A < \infty$ (by Theorem 1.51).||||

H. ANALYTICITY

A function $F : D \rightarrow B(X)$ (where D is a domain in $\mathbb{C}$) is *analytic* in D if

$$F'(z) := \lim_{h \to 0} h^{-1}[F(z + h) - F(z)]$$

exists in the uniform operator topology, for all $z \in D$. This is equivalent to the existence of that limit in the strong operator topology, and in the weak operator topology as well (cf. [HP, Theorem 3.10.1]).

1.53. DEFINITION. The C_o-semigroup $T(.)$ is *analytic* if it extends to an analytic function (also denoted $T(.)$) in some sector

$$S_\theta := \{z \in \mathbb{C}; |\arg z| < \theta, |z| > 0\},$$

$0 < \theta \leq \pi/2$, and $\lim T(z)x = x$ as $z \to 0, z \in S_\theta$.

The extended function necessarily satisfies the semigroup identity in S_θ:

$$T(z)T(w) = T(z + w) \qquad (z, w \in S_\theta)$$

(cf. [HP, Theorem 17.2.2]).

In the study of analyticity for C_o-semigroups, it may be assumed without loss of generality that the semigroup is uniformly bounded (consider $e^{-at}T(t)$ instead of $T(.)$). For simplicity, we consider only the special case of C_o-contraction semigroups, and the possibility of extending them as contraction-valued analytic semigroups. We refer to the literature for criteria applicable to the general case.

1.54. THEOREM. Let $T(.)$ be a C_o-semigroup of contractions, with generator A. Then $T(.)$ extends to an *analytic* contraction semigroup in a sector S_θ if and only if $e^{i\alpha} A$ generates a C_o-contraction semigroup for each $\alpha \in (-\theta, \theta)$.

PROOF. *Necessity.* For each $\alpha \in (-\theta, \theta)$, define $T_\alpha(t) := T(te^{i\alpha})$, $t \geq 0$.

Clearly, $T_\alpha(.)$ is a C_o-contraction semigroup. Denote its generator by A_α, and consider $\alpha > 0$ (the case $\alpha < 0$ is analogous). By Theorem 1.15, for all $s > 0$ and $x \in X$,

$$R(s; A_\alpha)x = \int_0^\infty e^{-st} T_\alpha(t)x \, dt$$

$$= e^{-i\alpha} \int_0^\infty \exp[-se^{-i\alpha}te^{i\alpha}]T(te^{i\alpha})x\, d(te^{i\alpha}). \tag{1}$$

The function $F(z) := \exp[-se^{-i\alpha}z]T(z)x$ is analytic in S_θ (for s, α, x fixed). Denote $C_a = \{z = ae^{i\phi}; 0 \le \phi \le \alpha\}$, oriented positively ($a > 0$). On C_a,

$$||F(z)|| \le ||x|| \exp[-s\Re(ze^{-i\alpha})] = ||x||e^{-sa\cos(\alpha-\phi)} \le ||x||e^{-sa\cos\alpha}.$$

Therefore the integral of F over C_a has norm $\le \pi ae^{-sa\cos\alpha}||x|| \to 0$ when $a \to 0+$ and when $a \to \infty$, since $0 < \alpha < \theta \le \pi/2$.

For $0 < a < b < \infty$, consider the closed contour

$$\Gamma_{a,b} := [a, b] + C_b - [a, b]e^{i\alpha} - C_a.$$

By Cauchy's theorem, $\int_{\Gamma_{a,b}} F(z)dz = 0$. Since the integrals of F on C_a and C_b converge strongly to 0 when $a \to 0+$ and $b \to \infty$, it follows that the right side of (1) is equal to $e^{-i\alpha} \int_0^\infty \exp[-(se^{-i\alpha})t]T(t)x\, dt$. However $\Re(se^{-i\alpha}) = s\cos\alpha > 0$, so that, by Theorem 1.15 (for the contraction case), the last expression is equal to $e^{-i\alpha}R(se^{-i\alpha}; A)x = R(s; e^{i\alpha}A)$. We conclude that A_α and $e^{i\alpha}A$ have equal resolvents on $\mathbb{R}^+$, and therefore $e^{i\alpha}A$ is indeed the generator of the C_o-contraction semigroup $T_\alpha(.)$.

Sufficiency. Suppose that for each $\alpha \in (-\theta, \theta)$, $e^{i\alpha}A$ generates a C_o- contraction semigroup $T_\alpha(.)$. By Theorem 1.36,

$$T_\alpha(t)x = \lim_n [\frac{n}{t}R(\frac{n}{t}; e^{i\alpha}A)]^n x$$

$$= \lim_n [I - \frac{z}{n}A]^{-n}x, \tag{2}$$

where $z = te^{i\alpha}, t > 0$.

Denote

$$F_n(z) := [I - \frac{z}{n}A]^{-n} = [\frac{n}{z}R(\frac{n}{z}; A)]^n.$$

Since A generates a C_o-contraction semigroup, F_n are analytic in $\Re\frac{n}{z} > 0$, hence in S_θ (if $z := te^{i\phi} \in S_\theta$, then $\Re(n/z) = (n/t)\cos\phi > 0$).

Since $F_n(z) = [\frac{n}{t}R(\frac{n}{t}; e^{i\alpha}A)]^n$, and $e^{i\alpha}A$ is the generator of a C_o-contraction semigroup, we have $||F_n(z)|| \le 1$ for all $z \in S_\theta$ (by Corollary 1.18).

For each $x \in X$ and $x^* \in X^*$, the sequence $\{x^*F_n(.)x\}$ of complex analytic functions is uniformly bounded (by $||x||.||x^*||$) in S_θ, hence is a normal family. It has then a subsequence converging uniformly on every compact subset of S_θ to a function $f(.; x, x^*)$ analytic in S_θ. By (2),

$$f(te^{i\alpha}; x, x^*) = x^*T_\alpha(t)x \tag{3}$$

for all $x \in X, x^* \in X^*, t > 0$, and $\alpha \in (-\theta, \theta)$.

Define $T(z) = T_\alpha(t)$, for $z = te^{i\alpha} \in S_\theta$. By (3), $T(.)$ is analytic in S_θ. It coincides with the original semigroup on $[0, \infty)$ and is contraction-valued in the sector (by definition). It remains to verify that

$$\lim \|T(z)x - x\| = 0$$

as $z \to 0$, $z \in S_\theta$ (for all $x \in X$). Since $\|T(.) - I\| \leq 2$ in the sector, we may consider only x in the dense set $D(A) = D(e^{i\alpha}A)$. For such x, writing $z = te^{i\alpha} \in S_\theta$, we have (since $e^{i\alpha}A$ generates the C_o-contraction semigroup $T_\alpha(t) = T(te^{i\alpha})$),

$$\|T(z)x - x\| = \|\int_0^t T_\alpha(s)e^{i\alpha}Ax\,ds\| \leq t\|Ax\| = |z|.\|Ax\|,$$

and the conclusion follows. ‖‖‖‖

COROLLARY 1. Let A generate a C_o-semigroup of contractions $T(.)$. Then $T(.)$ extends as an analytic semigroup of contractions in a sector S_θ $(0 < \theta \leq \pi/2)$ if and only if

$$\cos\alpha\Re(x^*Ax) - \sin\alpha\Im(x^*Ax) \leq 0 \qquad (*)$$

for all unit vectors $x \in D(A)$ and $x^* \in X^*$ such that $x^*x = 1$, and for all $\alpha \in (-\theta, \theta)$.

PROOF. For all $\alpha \in (-\theta, \theta)$, $e^{i\alpha}A$ is closed, densely defined, and for all $\lambda > 0$, $\lambda I - e^{i\alpha}A = e^{i\alpha}[\lambda e^{-i\alpha}I - A]$ is surjective, since $\Re(\lambda e^{-i\alpha}) = \lambda\cos\alpha > 0$ (cf. Theorem 1.26). Therefore, by Theorem 1.26, $e^{i\alpha}A$ generates a C_o-semigroup of contractions if and only if it is dissipative, i.e., if and only if

$$\Re(x^*e^{i\alpha}Ax) \leq 0$$

for all unit vectors $x \in D(A)$ and $x^* \in X^*$ such that $x^*x = 1$. This is precisely Condition $(*)$, so that the corollary follows immediately from Theorem 1.54.

When $\theta = \pi/2$ (i.e., for analytic semigroups in the right halfplane $\mathbb{C}^+$), we may consider "boundary values" on the imaginary axis.

1.55. THEOREM. Let $T(.)$ be an analytic semigroup in $\mathbb{C}^+$, and suppose it is *bounded* in the rectangle

$$Q := \{z = t + is \in \mathbb{C}; t \in (0, 1], s \in [-1, 1]\}.$$

Let $\nu := \log[\sup_Q \|T(.)\|]$ (of course, $0 \leq \nu < \infty$). Then for each $s \in \mathbb{R}$,

$$T(is) := \lim_{t \to 0+} T(t + is)$$

exists in $B(X)$ in the strong operator topology, and has the following properties:

(1) $T(i.)$ is a C_o-group;

(2) $T(is)$ commutes with $T(z)$ for all $s \in \mathbb{R}, z \in \mathbb{C}^+$;

(3) $T(t + is) = T(t)T(is)$ for all $t > 0, s \in \mathbb{R}$;

(4) $T(.)$ is of exponential type $\leq \nu$ in the *closed* right halfplane, i.e.,

$$\|T(z)\| \leq Ke^{\nu|z|} \qquad (Rez \geq 0);$$

and

(5) If A is the generator of $\{T(t); t \geq 0\}$, then iA is the generator of the **boundary group** $\{T(is); s \in \mathbb{R}\}$.

PROOF. See [HP], Theorem 17.9.1 and its proof.

COROLLARY 2. Suppose that the generator A of the C_o- semigroup of contractions $T(.)$ has *real* numerical range (i.e., $\nu(A) \subset \mathbb{R}$, cf. Definition 1.24). Then $T(.)$ extends as an analytic semigroup of contractions in $\mathbb{C}^+$. In particular, the boundary group $\{T(is); s \in \mathbb{R}\}$ exists, and is a C_o-group of isometries (with generator iA).

PROOF. Condition $(*)$ of the previous corollary reduces here to $\cos \alpha \; \Re(x^* A x) \leq 0$ (for all parameters in their proper ranges), which is trivially satisfied (since $|\alpha| < \theta \leq \pi/2$, and by Theorem 1.26 applied to A). Observe finally that a *group* of contractions consists in fact of isometries.

COROLLARY 3. Let A be a closed densely defined operator. Then the following are equivalent:

(1) A generates an analytic semigroup of contractions in the sector S_θ ($0 < \theta \leq \pi/2$);

(2) $\|zR(z; A)\| \leq 1$ for all $z \in S_\theta$;

(3) $zI - A$ is surjective for all $z \in S_\theta$, and $\Re[e^{i\alpha}\nu(A)] \leq 0$ for all $\alpha \in (-\theta, \theta)$.

PROOF. Writing $z = te^{i\alpha}$, we see that Condition (2) is equivalent to

(2') $\|tR(t; e^{i\alpha}A)\| \leq 1$ for all $t > 0$ and $\alpha \in (-\theta, \theta)$,

and (2') is equivalent to (1), by the Hille-Yosida Theorem (for contraction semigroups) and Theorem 1.54.

Assume (3). For all $\alpha \in (-\theta, \theta)$, $e^{i\alpha}A$ is closed, densely defined, and for all $t > 0$, $tI - e^{i\alpha}A = e^{i\alpha}[te^{-i\alpha}I - A]$ is surjective. The inequality in (3) means that $e^{i\alpha}A$ is dissipative, and (1) follows from Theorem 1.26 and Theorem 1.54. Conversely, if (1) holds, then Theorems 1.54 and 1.26 imply that $e^{i\alpha}A$ is dissipative and $tI - e^{i\alpha}A$ is surjective for all $\alpha \in (-\theta, \theta)$, and this is equivalent to (3).

COROLLARY 4. Let A be a closed densely defined operator such that $zI - A$ is surjective for $\Re z > 0$ and $\nu(A) \subset (-\infty, 0]$. Then A generates an analytic semigroup of contractions $T(.)$ in the right halfplane. In particular, the boundary group $\{T(is)\}$ exists, and is a C_o-group of isometries (with generator iA).

PROOF. For *real* numerical range, the inequality in Corollary 3 (3) reduces to $\cos \alpha \ x^* A x \leq 0$ (for all parameters in their proper ranges), which is satisfied by hypothesis. The conclusion follows then from Corollaries 2 and 3.

In case X is a *Hilbert* space, let $\pi : X^* \to X$ be the canonical isometric anti-isomorphism of X^* onto X given by the Riesz representation $x^* x = (x, \pi(x^*))$. Then

$$\nu(A) = \{(Ax, \pi(x^*)); x \in D(A), x^* \in X^*, ||x|| = ||x^*|| = (x, \pi(x^*)) = 1\}.$$

However, writing $\pi(x^*) = y$, we have (for x, x^* as in the above formula):

$$||x - y||^2 = ||x||^2 - 2\Re(x, y) + ||y||^2 = 1 - 2 + 1 = 0,$$

so that $y = x$, and

$$\nu(A) = \{(Ax, x); x \in D(A), ||x|| = 1\}.$$

Therefore A is *dissipative* if and only if

$$\Re(Ax, x) \leq 0 \qquad (x \in D(A)).$$

The inequality in Condition (3) of Corollary 1 becomes

$$\Re[e^{i\alpha}(Ax, x)] \leq 0 \qquad (x \in D(A)).$$

We then have

COROLLARY 5. Let A generate a C_o-semigroup of contractions $T(.)$ in Hilbert space. Then $T(.)$ extends as an analytic semigroup of contractions in a sector S_θ ($0 < \theta \leq \pi/2$) if and only if

$$\Re[e^{i\alpha}(Ax, x)] \leq 0 \qquad (x \in D(A), \alpha \in (-\theta, \theta)).$$

COROLLARY 6. Let A be a closed densely defined operator in Hilbert space, such that $zI - A$ is surjective for $\Re z > 0$ and $(Ax, x) \leq 0$ for all $x \in D(A)$. Then A generates an analytic semigroup of contractions in $\Re z > 0$; the boundary group $\{T(is)\}$ is a unitary C_o-group (with generator iA), and A is selfadjoint.

PROOF. By Corollary 2, A generates an analytic C_o-semigroup of contractions in $\Re z > 0$. Since (Ax, x) is *real* for all $x \in D(A)$, A is symmetric. The boundary group $\{T(is); s \in \mathbb{R}\}$ in that corollary (with generator iA) satisfies (cf. Theorem 1.36):

$$T(is)x = \lim_n [\frac{n}{s} R(\frac{n}{s}; iA)]^n x = \lim_n [\frac{n}{is} R(\frac{n}{is}; A)]^n x$$

for all $s > 0$ and $x \in X$. Since $-iA$ is the generator of the C_o-semigroup $\{T(-is); s \geq 0\}$, we also have

$$T(-is)x = \lim_n [\frac{n}{s} R(\frac{n}{s}; -iA)]^n x = \lim_n [\frac{n}{-is} R(\frac{n}{-is}; A)]^n x.$$

Since A is symmetric $R(z; A)^* = R(z^-; A)$, and in particular, $R(z; A)$ is normal. Therefore, for all $x, y \in X$ and $s > 0$,

$$(T(-is)x, y) = \lim_n([\frac{n}{is} R(\frac{n}{is}; A)]^{n*} x, y) = \lim_n(x, [\frac{n}{is} R(\frac{n}{is}; A)]^n y) = (x, T(is)y),$$

i.e.,

$$T(is)^* = T(-is) = T(is)^{-1}$$

for all $s > 0$ (hence for all $s \in \mathbb{R}$). Thus the boundary group is unitary; by Stone's theorem (Theorem 1.41), its generator iA has the form iH with H selfadjoint, that is, A is *selfadjoint*.||||

A more direct way to prove the selfadjointness of A goes as follows. Suppose $y \in X$ satisfies $((iI - A)x, y) = 0$ for all $x \in D(A)$. Then $i(x, y) = (Ax, y)$ for all $x \in D(A)$. Take $x = R(s; A)y \ (\in D(A)!)$ for some $s > 0$. Then

$$i(R(s; A)y, y) = (AR(s; A)y, y) = ([sR(s; A) - I]y, y).$$

The left side is pure imaginary, while the right side is real (since the bounded operators appearing there are both selfadjoint). Therefore

$$(R(s; A)y, y) = s(R(s; A)y, y) - (y, y) = 0,$$

hence $(y, y) = 0$ and $y = 0$. This shows that $iI - A$ (and similarly, $-iI - A$) has dense range, which means that A is *essentially selfadjoint* (cf. "digression" preceding Theorem 1.41). Since A is closed, it is actually *selfadjoint*.

Note that the relation $T(t)x = \lim_n [\frac{n}{t} R(\frac{n}{t}; A)]^n x$ shows that the operators $T(t)$ are *selfadjoint* (for A symmetric). The longer discussion given above illustrates the "method of analytic continuation to the imaginary axis", that will be used in Section 2.37 in the more general case of a "local semigroup" of *unbounded symmetric* operators to produce a selfadjoint operator H such that each $T(t)$ is a restriction of e^{-tH}.

K. NON-COMMUTATIVE TAYLOR FORMULA

In this section, we consider a C_o-semigroup $T(.)$ as a function of its generator A, when A varies in the set of all generators of C_o-semigroups. The notation $T(.) = T(.; A)$ will be used to exhibit the generator A of the semigroup.

In order to get a feeling about a possible Taylor formula relating $T(.; B)$ with $T(.; A)$ and *derivatives* of the semigroup with respect to A (at the *point* A), we consider first the case of uniformly continuous semigroups (i.e., the variable generator varies in $B(X)$). This case can be formulated in an arbitrary complex Banach algebra $\mathcal{A}$ with identity I, and we may consider analytic functions on it, more general than the functions $f_t(A) := e^{tA}$, $t \geq 0, A \in \mathcal{A}$. As before, we denote the resolvent set of an element $S \in \mathcal{A}$ by $\rho(S)$, etc... We start with the following elementary

1.56. LEMMA. Let $S, T \in \mathcal{A}$ and $z \in \rho(S) \cap \rho(T)$. Then for all $n = 0, 1, 2, ...,$

$$R(z; T) = \Sigma_{j=0}^n [R(z; S)(T - S)]^j R(z; S)$$

$$+ [R(z; S)(T - S)]^{n+1} R(z; T).$$

PROOF. If $Q \in \mathcal{A}$ is such that $I - Q$ is invertible in $\mathcal{A}$, then one verifies directly the "geometric series addition formula"

$$(I - Q)^{-1} = \Sigma_{j=0}^n Q^j + Q^{n+1}(I - Q)^{-1}, \tag{1}$$

$n = 0, 1, 2, ...$ For $z \in \rho(S) \cap \rho(T)$, we take

$$Q := R(z; S)(T - S) = R(z; S)[(zI - S) - (zI - T)]$$

$$= I - R(z; S)(zI - T),$$

so that $I - Q = R(z; S)(zI - T)$ is indeed invertible in $\mathcal{A}$ with inverse equal to $R(z; T)(zI - S)$. Substituting in (1) and multiplying on the right by $R(z; S)$, the lemma follows.||||

The formula of the lemma simplifies as follows when S, T are commuting elements of $\mathcal{A}$:

1.57. LEMMA. Let $S, T \in \mathcal{A}$ commute, and let $z \in \rho(S) \cap \rho(T)$. Then

$$R(z; T) = \Sigma_{j=0}^{n} R(z; S)^{j+1} (T - S)^j$$

$$+ R(z; S)^{n+1} R(z; T)(T - S)^{n+1},$$

$n = 0, 1, 2, \dots .$

Given *arbitrary* elements $A, B \in \mathcal{A}$, we consider the *commuting* multiplication operators $L_A, R_B \in B(\mathcal{A})$ defined by

$$L_A U = AU; \qquad R_B U = UB, \qquad (U \in \mathcal{A}).$$

We then have

1.58. LEMMA. Let $A, B \in \mathcal{A}$ and $z \in \rho(A) \cap \rho(B)$. Then for all $n = 0, 1, 2, \dots,$

$$R(z; B) = \Sigma_{j=0}^{n} R(z; A)^{j+1} (R_B - L_A)^j I$$

$$+ R(z; A)^{n+1} [(R_B - L_A)^{n+1} I] R(z; B).$$

and

$$R(z; B) = \Sigma_{j=0}^{n} [(L_B - R_A)^j I] R(z; A)^{j+1}$$

$$+ R(z; B)[(L_B - R_A)^{n+1} I] R(z; A)^{n+1}.$$

PROOF. We apply Lemma 1.57 to the commuting elements $S = L_A$ and $T = R_B$ of the Banach algebra $B(\mathcal{A})$. If $z \in \rho(A) \cap \rho(B)$, then $z \in \rho(L_A) \cap \rho(R_B)$, $R(z; L_A) = L_{R(z;A)}$, and $R(z; R_B) = R_{R(z;B)}$. Therefore

$$R_{R(z;B)} = \Sigma_{j=0}^{n} L_{R(z;A)}^{j+1} (R_B - L_A)^j + L_{R(z;A)}^{n+1} R_{R(z;B)} (R_B - L_A)^{n+1}.$$

Applying this operator to the identity $I \in \mathcal{A}$, we obtain the first formula of the lemma. The second formula is deduced in the same manner, through the choice $S = R_A$ and $T = L_B$ in Lemma 1.57.$\|\|\|$

The non-commutative Taylor formula for analytic functions on the Banach algebra $\mathcal{A}$ uses the Riesz-Dunford analytic operational calculus. Let f be a complex analytic function in an open neighborhood Ω of the spectrum $\sigma(B)$ of $B \in \mathcal{A}$. If $K \subset \Omega$ is compact, we denote by $\Gamma(K, \Omega)$ any finite union of positively oriented simple closed Jordan contours in Ω, that contains K in its interior. The element $f(B) \in \mathcal{A}$ is defined by

$$f(B) = \frac{1}{2\pi i} \int_{\Gamma} f(z) R(z; B) dz,$$

where $\Gamma = \Gamma(\sigma(B), \Omega)$, and the definition is independent of the choice of such Γ (cf. [DS I]).

1.59. THEOREM. Let $\mathcal{A}$ be a complex Banach algebra with identity I. Let $A, B \in \mathcal{A}$, and let f be a complex function analytic in a neighborhood Ω of $\sigma(A) \cup \sigma(B)$. Then for $n = 0, 1, 2, \ldots$,

$$f(B) = \Sigma_{j=0}^{n} \frac{f^{(j)}(A)}{j!}(R_B - L_A)^j I + L_n(f, A, B)$$

and

$$f(B) = \Sigma_{j=0}^{n}(L_B - R_A)^j I . \frac{f^{(j)}(A)}{j!} + R_n(f, A, B),$$

where the "left" and "right" remainders L_n and R_n are given by the formulas

$$L_n = \frac{1}{2\pi i} \int_\Gamma f(z)R(z; A)^{n+1}(R_B - L_A)^{n+1} I . R(z; B) dz$$

and

$$R_n = \frac{1}{2\pi i} \int_\Gamma f(z)R(z; B)(L_B - R_A)^{n+1} I . R(z; A)^{n+1} dz,$$

with $\Gamma = \Gamma(\sigma(A) \cup \sigma(B), \Omega)$.

PROOF. For Γ as above, the Riesz-Dunford operational calculus satisfies

$$f^{(j)}(A) = \frac{j!}{2\pi i} \int_\Gamma f(z)R(z; A)^{j+1} dz,$$

and the theorem follows from Lemma 1.58 by integration.||||

Note that

$$(R_B - L_A)^j I = \Sigma_{k=0}^{j} \binom{j}{k}(-A)^k B^{j-k},$$

with a similar formula for $(L_B - R_A)^j I$.

When A, B commute, these formulas reduce to $(B - A)^j$, and the Taylor formula of the theorem reduces to its "classical" form

$$f(B) = \Sigma_{j=0}^{n} \frac{f^{(j)}(A)}{j!}(B - A)^j$$

$$+ \frac{1}{2\pi i} \int_\Gamma f(z)R(z; A)^{n+1} R(z; B) dz . (B - A)^{n+1}.$$

When f is analytic in a "large" disc, the "Taylor formula" of Theorem 1.59 implies a non-commutative Taylor series expansion:

1.60. THEOREM. Let $\mathcal{A}$ be a complex Banach algebra with identity I, and let $A, B \in \mathcal{A}$. Suppose f is a complex function analytic on the *closed* disc

$$\{z \in \mathbb{C}; |z| \leq 2||A|| + ||B||\}.$$

Then

$$f(B) = \Sigma_{j=0}^{\infty} \frac{f^{(j)}(A)}{j!}(R_B - L_A)^j I = \Sigma_{j=0}^{\infty}(L_B - R_A)^j I . \frac{f^{(j)}(A)}{j!},$$

with both series converging strongly in $\mathcal{A}$.

PROOF. It suffices to prove that the remainders L_n and R_n converge strongly to 0 in $\mathcal{A}$.

Fix $r > 2||A|| + ||B||$ such that $C_r(:=$ the positively oriented circle of radius r centered at 0) and its interior are contained in the domain of analyticity Ω of f. Clearly $\sigma(A) \cup \sigma(B)$ lies in the interior of C_r, so we can take $\Gamma = C_r$ in Theorem 1.59. Let $M_r := \max_{z \in C_r} |f(z)|$.

On C_r, we have $||R(z; A)|| = ||\Sigma_{n=0}^{\infty} \frac{A^n}{z^{n+1}}|| \leq (r - ||A||)^{-1}$, and similarly for $R(z; B)$. Therefore

$$||L_n|| \leq \frac{rM_r}{(r - ||A||)^{n+1}(r - ||B||)}||(R_B - L_A)^{n+1}I||,$$

with a similar estimate for R_n (replace the last factor by $||(L_B - R_A)^{n+1}I||$).

However

$$||(R_B - L_A)^{n+1}I|| = ||\Sigma_{j=0}^{n+1}\binom{n+1}{j}(-A)^j B^{n+1-j}|| \leq (||A|| + ||B||)^{n+1},$$

and similarly for the letters R, L interchanged. Therefore

$$||L_n|| \leq \frac{rM_r}{r - ||B||}[\frac{||A|| + ||B||}{r - ||A||}]^{n+1} \to 0$$

as $n \to \infty$, because $||A|| + ||B|| < r - ||A||$, and similarly for R_n.||||

Taking $f(z) = f_t(z) := e^{tz}$ (for $t \geq 0$ fixed) in the "Taylor formula" of Theorem 1.59, we obtain

$$e^{tB} = e^{tA}\Sigma_{j=0}^{n}\frac{t^j}{j!}(R_B - L_A)^j I + L_n, \tag{1}$$

with the appropriate expression for the remainder L_n.

This is the "non-commutative Taylor formula" we wish to generalize to the case of strongly continuous semigroups.

For (generally) unbounded operators A, B, we use the (suggestive) notation

$$(B - A)^{[j]} := (R_B - L_A)^j I := \Sigma_{k=0}^j \binom{j}{k}(-A)^k B^{j-k}$$

with maximal domain

$$D((B - A)^{[j]}) = \bigcap_{k=0}^j D(A^k B^{j-k}).$$

The dense $T(.; A)$-invariant core for A consisting of all the C^∞-vectors for A (cf. Theorem 1.8) is denoted by $D^\infty(A)$. The *type* of $T(.; A)$ is $\omega(A)$ (cf. Section 1.3). We can state now

1.61. THEOREM. Let A, B be generators of C_o- semigroups such that $D^\infty(B) \subset D^\infty(A)$. Fix $a > \max[\omega(A), \omega(B)]$. Then for $n = 0, 1, 2, ...$ and $c > a$,

$$T(t; B)x = T(t; A)\Sigma_{j=0}^n \frac{t^j}{j!}(B - A)^{[j]}x + L_n(t; A, B)x$$

for all $x \in D^\infty(B)$ and $t \geq 0$, where the "n-th remainder" L_n is given by

$$L_n(t; A, B)x = \frac{1}{2\pi i} \int_{c-i\infty}^{c+i\infty} e^{tz} R(z; A)^{n+1}(B - A)^{[n+1]}R(z; B)x\,dz;$$

the integral converges strongly in X as a **Cauchy Principal Value** and is independent of $c > a$.

PROOF. Let A_s and B_u be the Hille-Yosida approximations of A and B respectively $(s, u > a$; cf. Sections 1.16 and 1.18).

We recall that there exists $M > 0$ and $r > a$ such that

$$||e^{tA_s}|| \leq Me^{at}; \qquad ||e^{tB_u}|| \leq Me^{at} \tag{2}$$

for all $s, u > r$ and $t \geq 0$. In particular, A_s and B_u have their spectra in the closed halfplane $\{z \in \mathbb{C}; \Re z \leq a\}$, and

$$||R(z; A_s)|| \leq \frac{M}{\Re z - a} \qquad (\Re z > a), \tag{3}$$

68

with a similar estimate for $R(z; B_u)$ (for all $s, u > r$).

Recall that, in the strong operator topology,

$$e^{tA_s} \to T(t; A) \qquad (t \geq 0), \tag{4}$$

and

$$R(z; A_s) \to R(z; A) \qquad (\Re z > a), \tag{5}$$

as $s \to \infty$. Also, for all $x \in D(A)$,

$$A_s x \to_{s \to \infty} Ax \tag{6}$$

(cf. Sections 1.16, 1.17, 1.32).

By (3), it follows from (5) that for all $j \in \mathbb{N}$ and $\Re z > a$,

$$R(z; A_s)^j \to R(z; A)^j \tag{7}$$

in the strong operator topology, as $s \to \infty$.

By (6), in the strong operator topology,

$$A_s R(z; A) \to_{s \to \infty} AR(z; A), \tag{8}$$

for each $z \in \rho(A)$.

By (3) and the definition of A_s,

$$\|A_s R(z; A)\| = \|sAR(s; A)R(z; A)\| = s\|R(s; A)[AR(z; A)]\|$$

$$\leq s\|R(s; A)\|.\|AR(z; A)\| \leq \frac{Ms}{s-a}\|AR(z; A)\| \leq 2M\|AR(z; A)\|$$

for all $s > a$. This uniform boundedness together with (8) imply that for all $m \in \mathbb{N}$,

$$[A_s R(z; A)]^m \to_{s \to \infty} [AR(z; A)]^m \tag{8'}$$

in the strong operator topology. Since

$$D(A^m) = R(z; A)^m X \qquad (m \in \mathbb{N}; z \in \rho(A)),$$

writing $x \in D(A^m)$ in the form $x = R(z; A)^m y$ for a suitable $y \in X$, we obtain (since A_s commutes with $R(z; A)$):

$$A_s^m x = A_s^m R(z; A)^m y = [A_s R(z; A)]^m y \to [AR(z; A)]^m y = A^m x.$$

Thus for all $m \in \mathbb{N}$,

$$A_s^m x \to_{s \to \infty} A^m x \qquad (x \in D(A^m)). \tag{9}$$

69

For $0 \leq k \leq j$, $x \in D(B^{j-k})$, and $s > r$ fixed, it follows from (9) that

$$(-A_s)^k B_u^{j-k} x \rightarrow (-A_s)^k B^{j-k} x$$

(as $u \rightarrow \infty$). Therefore

$$(B_u - A_s)^{[j]} x \rightarrow_{u \to \infty} (B - A_s)^{[j]} x \tag{10}$$

for all $x \in D(B^j)$. Hence

$$R(z; A_s)^{j+1}(B_u - A_s)^{[j]} x \rightarrow_{u \to \infty} R(z; A_s)^{j+1}(B - A_s)^{[j]} x \tag{11}$$

for all $x \in D(B^j)$, $\Re z > a$, $s > r$, and $j = 0, 1, \dots$.
   If $x \in D((B - A)^{[j]})$, then for $0 \leq k \leq j$, $B^{j-k} x \in D(A^k)$, and therefore (9) implies that $(-A_s)^k B^{j-k} x \rightarrow (-A)^k B^{j-k} x$ as $s \rightarrow \infty$, hence

$$(B - A_s)^{[j]} x \rightarrow_{s \to \infty} (B - A)^{[j]} x. \tag{12}$$

Together with (3) and (7), this implies that

$$R(z; A_s)^m (B - A_s)^{[j]} x \rightarrow_{s \to \infty} R(z; A)^m (B - A)^{[j]} x \tag{13}$$

for all $x \in D(B - A)^{[j]})$, $m \in \mathbb{N}$, and $\Re z > a$.
   If $x \in \bigcap_{j=0}^{n} D((B - A)^{[j]})$, then surely $x \in D(B^n)$, and it follows from (11) and (13) that

$$\lim_{s \to \infty} \lim_{u \to \infty} \Sigma_{j=0}^{n} R(z; A_s)^{j+1}(B_u - A_s)^{[j]} x = \Sigma_{j=0}^{n} R(z; A)^{j+1}(B - A)^{[j]} x \tag{14}$$

for $\Re z > a$.
   On the other hand, by Lemma 1.58, for all $x \in X$, the left hand side of (14) is equal to

$$R(z; B_u)x - R(z; A_s)^{n+1}(B_u - A_s)^{[n+1]} R(z; B_u)x. \tag{15}$$

If $x \in D(B^{m-1})(= R(z; B)^{m-1} X)$ (for any $m \in \mathbb{N}$), writing $x = R(z; B)^{m-1} y$ for a suitable $y \in X$, we have

$$B_u^m R(z; B_u)x = [B_u R(z; B_u)][B_u R(z; B)]^{m-1} y.$$

The operators in the first bracket on the right are equal to $zR(z; B_u) - I$, and are therefore uniformly bounded (with respect to u) by $1 + \frac{M|z|}{\Re z - a}$ (by (3)), and converge (as $u \rightarrow \infty$) to $zR(z; B) - I = BR(z; B)$ (by (5)) in the strong operator topology. The second bracket converges to $[BR(z; B)]^{m-1} y$, by (8') for B. It follows that

$$B_u^m R(z; B_u)x \rightarrow [BR(z; B)][BR(z; B)]^{m-1} y = B^m R(z; B)^m y = B^m R(z; B)x$$

for all $x \in D(B^{m-1})$ and $\Re z > a$.

Therefore, for $s > r$ fixed and $x \in D(B^n)$, the right hand side of (15), which is equal to

$$R(z; B_u)x - \Sigma_{k=0}^{n+1} \binom{n+1}{k} R(z; A_s)^{n+1}(-A_s)^k B_u^{n+1-k} R(z; B_u)x,$$

converges as $u \to \infty$ to

$$R(z; B)x - R(z; A_s)^{n+1}(B - A_s)^{[n+1]} R(z; B)x$$

(cf. (5)).

If $x \in D(B^n)$ is such that $R(z; B)x \in D((B - A)^{[n+1]})$, it follows from (13) that the last expression converges to

$$R(z; B)x - R(z; A)^{n+1}(B - A)^{[n+1]} R(z; B)x$$

as $s \to \infty$.

We then conclude from (14) that the following generalization of Lemma 1.58 (first formula) is valid:

LEMMA 1. Let A, B be generators of C_o-semigroups. Then for $\Re z > \max[\omega(A), \omega(B)]$,

$$R(z; B)x = \Sigma_{j=0}^{n} R(z; A)^{j+1}(B - A)^{[j]} x + R(z; A)^{n+1}(B - A)^{[n+1]} R(z; B)x$$

for all $x \in \bigcap_{j=0}^{n} D((B - A)^{[j]})$ such that $R(z; B)x \in D((B - A)^{[n+1]})$ (i.e., for all x in the maximal domain of the right-hand side).

Assume now that $D^\infty(B) \subset D^\infty(A)$. If $0 \le k \le j$ and $x \in D^\infty(B)$, then $B^{j-k}x \in D^\infty(B) \subset D^\infty(A) \subset D(A^k)$, so that $x \in D((-A)^k B^{j-k})$. Hence $x \in \bigcap_{j=0}^{n} D((B - A)^{[j]})$ for all n. Since also $R(z; B)x \in D^\infty(B)$, we have $R(z; B)x \in D((B - A)^{[n+1]})$ as well, and the formula in the lemma is valid for all $x \in D^\infty(B)$.

We need the following generalization of the second formula in Theorem 1.15.

LEMMA 2. Let A generate the C_o-semigroup $T(.; A)$. Then for $t > 0$, $c > \omega(A)$, and $j = 0, 1, 2, ...$,

$$\lim_{\tau \to \infty} \frac{1}{2\pi i} \int_{c-i\tau}^{c+i\tau} e^{tz} R(z; A)^{j+1} x \, dz = \frac{t^j}{j!} T(t; A)x \qquad (x \in D(A)).$$

71

PROOF (of Lemma 2). Since $R(z; A)^{j+1} = \frac{(-1)^j}{j!} R(z; A)^{(j)}$, we may integrate by parts j times to show that the integral appearing in the lemma is equal to

$$[-e^{tz} \Sigma_{k=0}^{j-1} \frac{(j-k-1)!}{j!} t^k R(z; A)^{j-k} x]_{c-i\tau}^{c+i\tau}$$

$$+ \frac{t^j}{j!} \int_{c-i\tau}^{c+i\tau} e^{tz} R(z; A) x \, dz. \tag{16}$$

The "integrated part" has norm

$$\leq e^{ct} \Sigma_{k=0}^{j-1} t^k (||R(c+i\tau; A)^{j-k} x|| + ||R(c-i\tau; A)^{j-k} x||).$$

If $x \in D(A)$, write $x = R(\lambda; A) y$ for some λ with $\Re\lambda > a$. Then since $j - k \geq 1$,

$$||R(c+i\tau; A)^{j-k} x|| = ||R(c+i\tau; A)^{j-k-1} \frac{R(\lambda; A) y - R(c+i\tau; A) y}{c+i\tau - \lambda}||$$

$$\leq \frac{M^{j-k} ||y||}{(c-a)^{j-k-1} |c+i\tau - \lambda|} (\frac{1}{c-a} + \frac{1}{\Re\lambda - a})$$

$$\to_{\tau \to \infty} 0,$$

and similarly for $c - i\tau$. Therefore the integrated part in (16) converges to 0 when $\tau \to \infty$. By Theorem 1.15, the integral in (16) converges to $2\pi i T(t; A) x$ (for $x \in D(A)$), and the lemma follows.

If $x \in D^\infty(B)$ and $t > 0$, we have by Lemma 1

$$\int_{c-i\tau}^{c+i\tau} e^{tz} R(z; A)^{n+1} (B - A)^{[n+1]} R(z; B) x \, dz$$

$$= \int_{c-i\tau}^{c+i\tau} e^{tz} R(z; B) x \, dz - \Sigma_{j=0}^{n} \int_{c-i\tau}^{c+i\tau} e^{tz} R(z; A)^{j+1} (B - A)^{[j]} x \, dz. \tag{17}$$

However, for $x \in D^\infty(B)$, we surely have $x \in D(B)$, so that the first term on the right converges to $2\pi i T(t; B) x$, by Theorem 1.15 (when $\tau \to \infty$). We observed above that $B^{j-k} x \in D^\infty(A)$ for all $0 \leq k \leq j$, and therefore $(-A)^k B^{j-k} x \in D^\infty(A)$, and so $(B - A)^{[j]} x \in D^\infty(A) \subset D(A)$. Hence, by Lemma 2, the sum on the right of (17) converges to

$$2\pi i T(t; A) \Sigma_{j=0}^{n} \frac{t^j}{j!} (B - A)^{[j]} x. \tag{18}$$

This shows that the remainder L_n in Theorem 1.61 converges (as a "Cauchy Principal Value"), and its "value" is independent of $c > a$, and is equal to

$$T(t; B) - T(t; A) \Sigma_{j=0}^{n} \frac{t^j}{j!} (B - A)^{[j]} x. ||||$$

PART II. GENERALIZATIONS

A. PRE-SEMIGROUPS

We consider the following elementary properties of a C_o- semigroups $S(.)$:

Property 1. $S(.) : [0, \infty) \to B(X)$ is strongly continuous and $S(0)$ is injective.

Property 2. $S(t - u)S(u)$ is independent of u, for all $0 \le u \le t$.

Property 3. There exists $a \ge 0$ such that $e^{-at}S(t)x$ is bounded and uniformly continuous on $[0, \infty)$, for each $x \in X$.

Property 1. is contained in Theorem 1.1 (together with the trivial injectivity of $S(0) = I$). Property 2. follows from the semigroup identity. Property 3. follows from Theorem 1.1 and the estimate

$$||e^{-a(t+h)}S(t+h)x - e^{-at}S(t)x||$$

$$\le e^{-at}||S(t)||.||e^{-ah}S(h)x - x|| \le M||e^{-ah}S(h)x - x||.$$

2.1. DEFINITION. A **pre-semigroup** is a function $S(.)$ with the properties 1. and 2. If Property 3. is also satisfied, the pre-semigroup is **exponentially tamed**.

By 2., equating the values of $S(t - u)S(u)$ with the value at $u = t$, we see that

$$S(t - u)S(u) = S(0)S(t) \qquad (t \ge u \ge 0). \tag{1}$$

Writing $t = u + s$ in (1) , the identity is equivalent to

$$S(s)S(u) = S(0)S(u + s) \qquad (s, u \ge 0). \tag{1'}$$

In particular, the values of $S(.)$ commute.

2.2. DEFINITION. The **generator** A of the pre-semigroup $S(.)$ has domain $D(A)$ consisting of all $x \in X$ for which the strong right derivative at 0, $[S(.)x]'(0)$, exists and belongs to $S(0)X$, and

$$Ax = S(0)^{-1}[S(.)x]'(0) \qquad (x \in D(A)).$$

Note that if $T(.)$ is a C_o-semigroup with generator A, and $C \in B(X)$ is injective and commutes with $T(.)$, then $S(.) := CT(.)$ is a pre-semigroup with $S(0) = C$ and with generator A.

We first generalize Theorem 1.2 as follows

2.3. THEOREM. Let A generate the pre-semigroup $S(.)$. Then:
1. A commutes with $S(t)$ for all $t \geq 0$.
2. A is closed with $S(0)X \subset D(A)^-$.
3. For each $x \in D(A)$, $u := S(.)x$ is of class C^1 and solves

$$(ACP) \qquad u' = Au; \qquad u(0) = S(0)x$$

on $[0, \infty)$.

PROOF. For $t \geq 0$, $h > 0$, and $x \in D(A)$,

$$S(h)[S(t)x] - S(0)[S(t)x] = S(t)[S(h)x - S(0)x] = S(0)S(t+h)x - S(0)S(t)x.$$

Dividing by h and letting $h \to 0$, we get that the strong right derivative at 0 of $S(.)[S(t)x]$ exists, equals the strong right derivative of $S(0)S(.)x$ at t, and equals $S(t)S(0)Ax = S(0)S(t)Ax \in S(0)X$. Therefore $S(t)x \in D(A)$ and $A[S(t)x] := S(0)^{-1}[S(0)S(t)Ax] = S(t)Ax$. This proves 1.

Also for $0 < h \leq t$ (with t fixed), letting $K := \sup_{0 < u \leq t} ||S(u)||(< \infty$ by the uniform boundedness theorem and Property 1.), we have

$$||h^{-1}[S(0)S(t-h)x - S(0)S(t)x] + S(0)S(t)Ax||$$

$$\leq ||h^{-1}[S(0)S(t-h) - S(0)S(t)]S(h)x + S(0)S(t)Ax||$$

$$+||S(0)||.||S(t-h) - S(t)||.||h^{-1}[S(h)x - x] - S(0)Ax||$$

$$+||S(0)||.||S(t)[S(0)Ax] - S(t-h)[S(0)Ax]||.$$

The first term on the right of the inequality equals

$$||S(0)\{-h^{-1}[S(0)S(t+h) - S(0)S(t)] + S(t)Ax\}|| \to 0$$

when $h \to 0$, as observed before.
    The second term on the right is

$$\leq 2K||S(0)||.||h^{-1}[S(h)x - x] - S(0)Ax|| \to 0$$

by definition of A.
    The third term on the right $\to 0$ by continuity of $S(.)[S(0)Ax]$.

Thus we proved that $S(0)S(.)x$ is differentiable on $[0, \infty)$ for each $x \in D(A)$, and

$$[S(0)S(.)x]'(t) = S(0)S(t)Ax \qquad (t \geq 0; x \in D(A)). \qquad (2)$$

Integrating from 0 to t and using the boundedness and injectivity of $S(0)$, we obtain

$$S(t)x - S(0)x = \int_0^t S(s)Ax\,ds = \int_0^t AS(s)x\,ds \qquad (x \in D(A)). \qquad (3)$$

For $h, t > 0$ and for all $x \in X$, we have

$$h^{-1}[S(h) - S(0)] \int_0^t S(s)x\,ds = S(0)h^{-1}[\int_0^t S(s+h)x\,ds - \int_0^t S(s)x\,ds]$$

$$= S(0)[h^{-1} \int_t^{t+h} S(s)x\,ds - h^{-1} \int_0^h S(s)x\,ds] \to S(0)[S(t)x - S(0)x] \in S(0)X,$$

showing that $\int_0^t S(s)x\,ds \in D(A)$ and

$$A \int_0^t S(s)x\,ds = S(t)x - S(0)x \qquad (x \in X). \qquad (4)$$

In particular, for all $x \in X$,

$$S(0)x = \lim_{t \to 0+} t^{-1} \int_0^t S(s)x\,ds \in D(A)^-.$$

We show now that A is closed. If $x_n \in D(A)$, $x_n \to x$, and $Ax_n \to y$, then with K as before (for t fixed) and $L = \sup_n ||Ax_n||$, we have $||S(s)Ax_n|| \leq KL$ for all n and $s \in [0, t]$, and $S(s)Ax_n \to S(s)y$ pointwise. By dominated convergence and (3),

$$S(t)x - S(0)x = \lim_n [S(t)x_n - S(0)x_n] = \lim_n \int_0^t S(s)Ax_n\,ds = \int_0^t S(s)y\,ds.$$

Dividing by t and letting $t \to 0+$, we obtain that the right hand side converges to $S(0)y \in S(0)X$, so that $x \in D(A)$ and $Ax = y$, as wanted.

We read also from (3) that $S(.)x$ is of class C^1 and solves ACP on $[0, \infty)$ for each $x \in D(A)$.||||

A partial converse of Theorem 2.3 is the following

2.4. THEOREM. Let $S(.)$ have Property 1. and commute with A, and either $D(A)$ is dense or $\rho(A)$ is non-empty. If $S(.)x$ solves ACP for each $x \in D(A)$, then $S(.)$ is a pre-semigroup generated by an extension of A.

77

PROOF. For $x \in D(A)$ and $0 \le u \le t$,

$$\frac{d}{du}S(t-u)S(u)x = -AS(t-u)S(u)x + S(t-u)AS(u)x = 0,$$

and Property 2. follows on $D(A)$, hence on X in case $D(A)$ is dense, because $S(t-u)S(u) \in B(X)$.

In case $\rho(A)$ is non-empty, fix $\lambda \in \rho(A)$. Since $R(\lambda; A)x \in D(A)$ for all $x \in X$, and since $R(\lambda; A)$ commutes with $S(.)$ (because A commutes with $S(.)$), we have

$$R(\lambda; A)S(t-u)S(u)x = S(t-u)S(u)R(\lambda; A)x$$

$$= S(0)S(t)R(\lambda; A)x = R(\lambda; A)S(0)S(t)x,$$

and therefore $S(t-u)S(u) = S(0)S(t)$, i.e., Property 2. is satisfied.

Let then A' be the generator of $S(.)$. By hypothesis, $[S(.)x]'(0) = A[S(0)x] = S(0)Ax \in S(0)X$ for all $x \in D(A)$, that is, $A \subset A'$.||||

A generalization of Theorem 2.4 is the following

2.5. THEOREM. Let A, B be (unbounded) operators such that
(i) $0 \in \rho(B)$;
(ii) $D(B) \subset D(A)$; and
(iii) B commutes with $R(\lambda; A)$ for some $\lambda > 0$.
Then ACP for A has a unique C^1-solution on $[0, \infty)$ for each $x \in D(B)$ if and only if an extension of A generates a pre-semigroup $S(.)$ (with $S(0) = (\lambda I - A)R(0; B)$) that commutes with A.

PROOF. Suppose $S(.)$ is a pre-semigroup with $S(0)$ as stated, commuting with A and generated by an extension A' of A. First, $S(t)D(A) = S(t)R(\lambda; A)X = R(\lambda; A)S(t)X \subset D(A)$ for all t. For $x = S(0)y$ with $y \in D(A) \subset D(A')$, we have by Theorem 2.3

$$[S(.)y]' = A'[S(.)y] = A[S(.)y],$$

i.e., $[S(.)S(0)^{-1}x]' = A[S(.)S(0)^{-1}x]$ and of course $[S(.)S(0)^{-1}x](0) = x$, that is, $S(.)S(0)^{-1}x$ solves ACP for $x \in S(0)D(A) = D(B)$, since

$$S(0)D(A) = S(0)R(\lambda; A)X = R(\lambda; A)S(0)X = R(0; B)X = D(B).$$

If $v : [0, \infty) \to D(A)$ is any solution of ACP with $x = S(0)y \in S(0)D(A)$, then $\frac{d}{ds}[S(t-s)v(s)] = -S(t-s)A'v(s) + S(t-s)Av(s) = 0$ since $v(s) \in D(A)$. Equating therefore the values of the constant s-function $S(t-s)v(s)$ at $s = 0$ and $s = t$, we get $S(0)v(t) = S(t)S(0)y$, hence $v(t) = S(t)y = S(t)S(0)^{-1}x$, meaning that ACP has a unique solution for each $x \in S(0)D(A)$.

Conversely, assume ACP with initial value $x \in D(B)$ has the unique C^1- solution $u(.; x)$ on $[0, \infty)$. If $v := R(\lambda; A)u(.; x)$, then

$$v' = R(\lambda; A)u(.; x)' = R(\lambda; A)Au(.; x) = Av,$$

and

$$v(0) = R(\lambda; A)x = R(\lambda; A)R(0; B)y = R(0; B)R(\lambda; A)y \in D(B).$$

By the uniqueness assumption,

$$R(\lambda; A)u(.; x) = u(.; R(\lambda; A)x) \qquad (x \in D(B)). \qquad (1)$$

We define now for all $x \in X$

$$S(.)x := (\lambda I - A)u(.; R(0; B)x) = \lambda u(.; R(0; B)x) - u'(.; R(0; B)x). \qquad (2)$$

Since $R(0; B)x \in D(B)$, and $u(.; y)$ has values in $D(A)$ for any $y \in D(B)$, the operator $S(t)$ is everywhere defined on X, and is linear by the uniqueness hypothesis (for each $t \geq 0$). By (2), $S(.)x$ is continuous for each $x \in X$.

By (1) with $R(0; B)x (\in D(B))$ replacing x,

$$R(\lambda; A)S(t)x = u(t; R(0; B)x) = (\lambda I - A)R(\lambda; A)u(t; R(0; B)x)$$

$$= (\lambda I - A)u(t; R(\lambda; A)R(0; B)x) = (\lambda I - A)u(t; R(0; B)R(\lambda; A)x) = S(t)R(\lambda; A)x,$$

and it follows that $S(t)$ commutes with A for all t.

Consider now $U(.)x := u(.; R(0; B)x)$.

The operator $U(.) : X \to C^1([0, b]; X)$ into the Banach space of all X-valued C^1-functions on $[0, b]$ with the usual norm, is shown to be closed. Indeed, if $x_n \to x$ in X, and $U(.)x_n \to v$ in $C^1([0, b]; X)$, then for each $t \in [0, b]$,

$$A[U(t)x_n] = Au(t; R(0; B)x_n) = [u(.; R(0; B)x_n]'(t)$$

$$= [U(.)x_n]'(t) \to v'(t).$$

Since A is closed, $v(t) \in D(A)$ and $Av(t) = v'(t)$. Also

$$v(0) = \lim_n U(0)x_n = \lim_n u(0; R(0; B)x_n)$$

$$= \lim_n R(0; B)x_n = R(0; B)x.$$

By uniqueness, it follows that $v = u(.; R(0; B)x) = U(.)x$, that is $U(.)$ is closed, hence bounded, by the Closed Graph Theorem. Let M denote its norm. Then for $0 \leq t \leq b$,

$$||S(t)x|| = ||(\lambda I - A)U(t)x|| \leq \lambda||U(t)x|| + ||[U(.)x]'(t)||$$

$$\leq (\lambda + 1)M||x|| \qquad (x \in X).$$

Since b is arbitrary, this shows that $S(.)$ is $B(X)$-valued. We saw that it satisfies Property 1. with $S(0) = (\lambda I - A)R(0; B)x$ clearly injective, and it commutes with A. For $x \in D(A)$, write $x = R(\lambda; A)y$; then by (1),

$$S(.)x = (\lambda I - A)u(.; R(0; B)R(\lambda; A)y) = u(.; R(0; B)y)$$

solves ACP (with the initial value $S(0)x = R(0; B)y$).

By Theorem 2.4, we conclude that $S(.)$ is a pre-semigroup generated by an extension of A. ||||

Taking $B = -(\lambda I - A)^{n+1}$ (for some non-negative integer n) with domain $D(B) = D(A^{n+1}) \subset D(A)$, the pre-semigroup $S(.)$ generated by an extension of A satisfies $S(0) = (\lambda I - A)(\lambda I - A)^{-n-1} = R(\lambda; A)^n$. We thus have the following

2.6. COROLLARY. Let $\lambda \in \rho(A)$. Then ACP for A has a unique C^1-solution on $[0, \infty)$ for each $x \in D(A^{n+1})$ if and only if an extension of A generates a pre-semigroup $S(.)$ with $S(0) = R(\lambda; A)^n$, which commutes with A.

We consider next an *exponentially tamed* pre-semigroup $S(.)$, that is, all three properties 1.,2.,3. are satisfied.

By the injectivity of $S(0)$ and the Uniform Boundedness Theorem,

$$0 < ||S(0)|| \leq M := \sup_{t \geq 0} e^{-at}||S(t)|| < \infty.$$

2.7. DEFINITION. Let $S(.)$ be a pre-semigroup with Property 3. Set

$$Y = \{x \in X; S(0)^{-1}e^{-at}S(t)x \in C_b([0, \infty); X)\},$$

where $C_b(...)$ denotes the Banach space of all X-valued bounded uniformly continuous functions on $[0, \infty)$, normed by $||f||_u = \sup_{t \geq 0} ||f(t)||$.

For $x \in Y$, set
$$||x||_Y := ||S(0)^{-1}e^{-at}S(t)x||_u.$$

Clearly $||.||_Y \geq ||.||$ on Y, and $Y := (Y, ||.||_Y)$ is a normed space.

Finally, we denote by $[S(0)X]$ the Banach space $S(0)X$ with the norm

$$||x||_0 := M||S(0)^{-1}x||.$$

2.8. THEOREM. Let A generate an exponentially tamed pre-semigroup $S(.)$, and let Y be the space defined in 2.7. Then Y is a Banach subspace of X containing $[S(0)X]$ as a Banach subspace, and A_Y (the part of A in Y) generates a C_o-semigroup $T(.)$ in Y satisfying $||T(t)||_{B(Y)} \leq e^{at}$.

PROOF. Let $\{x_n\}$ be Cauchy in Y. Since $||.||_Y \geq ||.||$, it is Cauchy in X. Let then $x = \lim x_n$ in X. By definition of the Y-norm, the sequence $\{S(0)^{-1}e^{-at}S(t)x_n\}$ is Cauchy in $C_b := C_b([0,\infty); X)$. Let $u \in C_b$ be its C_b-limit. The boundedness of $S(0)$ implies that $e^{-at}S(t)x = S(0)u(t) \in S(0)X$ for all t, so that $S(0)^{-1}e^{-at}S(t)x = u(t) \in C_b$, i.e., $x \in Y$, and clearly $||x_n - x||_Y \to 0$. Thus Y is indeed a Banach subspace of X.

If $x = S(0)y \in S(0)X$, then $S(0)^{-1}e^{-at}S(t)x = e^{-at}S(t)y \in C_b$ by Property 3., so that $S(0)X \subset Y$. Also

$$||x||_Y \leq ||e^{-at}S(t)y||_u \leq M||y|| = ||x||_0.$$

Hence $[S(0)X]$ is a Banach subspace of Y.

If $x \in Y$, then for each fixed $s \geq 0$, $S(0)^{-1}e^{-at}S(t)S(s)x = e^{-at}S(t+s)x \in C_b$, (as a function of t), so that $S(s)Y \subset Y$. We may then define $T(.) := S(0)^{-1}S(.)$ on Y. Then for $x \in Y$,

$$||T(t)x||_Y = \sup_s ||S(0)^{-1}e^{-as}S(s)S(0)^{-1}S(t)x||$$

$$= e^{at} \sup_s ||S(0)^{-1}e^{-a(s+t)}S(s+t)x|| \leq e^{at}||x||_Y.$$

This shows that $T(.)$ is $B(Y)$-valued and $||T(t)||_{B(Y)} \leq e^{at}$.

The semigroup property of $T(.)$ follows trivially from (1').

The C_o-property of $T(.)$ follows from the uniform continuity of $S(0)^{-1}e^{-at}S(t)x$ for $x \in Y$. Indeed, given $\epsilon > 0$, there exists $\delta > 0$ such that for $0 < h < \delta$,

$$||S(0)^{-1}e^{-a(t+h)}S(t+h)x - S(0)^{-1}e^{-at}S(t)x||_u < \epsilon.$$

Therefore

$$||S(0)^{-1}e^{-at}S(t)[T(h)x - x]|| \leq$$

$$e^{ah}||S(0)^{-1}e^{-a(t+h)}S(t+h)x - S(0)^{-1}e^{-at}S(t)x||$$

$$+(e^{ah} - 1)||S(0)^{-1}e^{-at}S(t)x|| \leq e^{ah}\epsilon + (e^{ah} - 1)||x||_Y.$$

Taking the supremum over all $t \geq 0$, and letting then $h \to 0$, we obtain

$$\limsup_{h \to 0+} ||T(h)x - x||_Y \leq \epsilon,$$

and the arbitrariness of ϵ gives the C_o-property of $T(.)$ in Y.

Let then A' be the generator of $T(.)$ in Y. For $x \in D(A') \subset Y$ and $h > 0$,

$$h^{-1}[S(h)x - S(0)x] = S(0)h^{-1}[T(h)x - x] \to S(0)A'x \in S(0)X,$$

where the limit (as $h \to 0$ is in Y, hence in X). Therefore $x \in D(A)$ and $Ax = A'x \in Y$, by definition. This shows that $x \in D(A_Y)$ and $A_Y x = A'x$, i.e., $A' \subset A_Y$.

The Laplace transform $L(\lambda)x$ of $S(.)x$ is well-defined for $\lambda > a$, and calculations identical with those in the proof of Theorem 1.15 show that

$$L(\lambda)(\lambda I - A)x = S(0)x \qquad (x \in D(A)).$$

If $(\lambda I - A)x = 0$, it follows that $S(0)x = 0$, hence $x = 0$, i.e., $(\lambda I - A)$ is injective on $D(A)$, and therefore $(\lambda I - A_Y)$ is injective (for $\lambda > a$). Also, by Theorem 1.15, $R(\lambda; A') \in B(Y)$, so that in particular $\lambda I - A'$ is surjective (for those λ), and we saw above that $\lambda I - A' \subset \lambda I - A_Y$. It follows that $D(A') = D(A_Y)$.||||

B. SEMI-SIMPLICITY MANIFOLD (real spectrum case)

In this section, we generalize Theorem 1.49 to operators A with real spectrum, for which iA is *not* assumed to generate a C_o-group.

Consider the **Poissonian** of A,

$$P(t,s) := \frac{1}{2\pi i}[R(t - is; A) - R(t + is; A)] \qquad (t \in \mathbb{R}; s > 0).$$

Let $||.||_1$ denote the $L^1(\mathbb{R})$-norm (with respect to the Lebesgue measure).

2.9. DEFINITION. The **semi-simplicity manifold** for A is the set of all $x \in X$ such that

(1) $\lim_{|u| \to \infty} R(t + iu; A)x = 0$ for all $t \in \mathbb{R}$; and
(2) $\sup_{s > 0} ||x^* P(., s)x||_1 < \infty$ for all $x^* \in X^*$.

Note that Condition (1) is valid for all $x \in X$ when iA generates a C_o-group; indeed, since $R(t + iu; A) = iR(-u + it; iA)$, we have in that case $||R(t + iu; A)|| = O(1/|u|)$ for $0 \neq u \in \mathbb{R}$.

2.10. LEMMA. If $x \in X$ satifies Condition (2), then

$$||x||_A := \sup\{||x^* P(., s)x||_1 ; s > 0, ||x^*|| = 1\} < \infty.$$

PROOF. First, for $s > 0$ and $x \in X$ fixed, *assume* that $x^* P(., s)x \in L^1(\mathbb{R})$ for all $x^* \in X^*$, and consider then the linear map

$$V_s : x^* \to x^* P(., s)x$$

of X^* into $L^1(\mathbb{R})$.

If $x_n^* \to x^*$ in X^* and $V_s x_n^* \to f$ in L^1, then by Fatou's lemma,

$$\int_{\mathbb{R}} |f(t) - x^* P(t, s)x| dt = \int_{\mathbb{R}} \liminf_n |f(t) - x_n^* P(t, s)x| dt$$

$$\leq \liminf_n \int |f(t) - V_s x_n^*| dt = 0,$$

i.e., $V_s x^* = f$, so that V_s is closed, hence bounded, by the Closed Graph Theorem. When Condition (2) is satisfied by x, the family of *bounded* operators V_s satisfies

$$\sup_{s>0} ||V_s x^*||_1 < \infty \qquad (x^* \in X^*).$$

By the Uniform Boundedness Theorem, $\sup_{s>0} ||V_s|| < \infty$, which means precisely that $||x||_A < \infty$. ||||

2.11. THEOREM. Let A be an operator with real spectrum acting in the reflexive Banach space X, and let Z be its semi-simplicity manifold, normed by $||.||_A$. Then Z is a Banach subspace of X, invariant for any $U \in B(X)$ commuting with A, and there exists a spectral measure on Z, $E(.)$, such that

1. for each $\delta \in \mathcal{B}(\mathbb{R})$, $E(\delta)$ commutes with every $U \in B(X)$ which commutes with A;
2. $D(A_Z) = \{x \in Z; \int_{\mathbb{R}} uE(du)x$ exists and belongs to $Z\}$, and

$$Ax = \int_{\mathbb{R}} uE(du)x \qquad (x \in D(A_Z));$$

3. for all non-real $\zeta \in \mathbb{C}$ and $x \in Z$,

$$R(\zeta; A)x = \int_{\mathbb{R}} \frac{1}{\zeta - u} E(du)x.$$

Moreover, Z is "maximal-unique" in the following sense: if W is a linear manifold in X and $F(.)$ is a spectral measure on W with Property 3., then $W \subset Z$ and $F(\delta) = E(\delta)/W$ for all $\delta \in \mathcal{B}(\mathbb{R})$.

Note that the "existence" of the integral in 2. is in the sense of Section 1.48, i.e., as the strong limit in X

$$\int_{\mathbb{R}} uE(du)x := lim_{a,b} \int_a^b uE(du)x.$$

PROOF. For fixed $x \in Z$ and $x^* \in X^*$, the function $x^*P(t,s)x$ is an analytic function of $t + is$ in $\mathbb{C}^+$, hence (complex) harmonic there, and satisfies

$$\sup_{s>0} ||x^*P(.,s)x||_1 \leq ||x||_A||x^*|| < \infty$$

(by Lemma 2.10).

Therefore there exists a unique regular complex Borel measure $\mu(.; x, x^*)$ on $\mathcal{B}(\mathbb{R})$ such that

$$||\mu(.; x, x^*)|| \leq ||x||_A||x^*|| \qquad (1)$$

84

and

$$x^* P(t,s) x = \int_{\mathbb{R}} p(t - u, s) \mu(du; x, x^*) \quad (t \in \mathbb{R}, s > 0), \tag{2}$$

where $p(t,s) := \frac{s}{\pi(t^2 + s^2)}$ is the Poisson kernel for the upper half-plane (cf. [SW, pp. 49-53]). The uniqueness of the Poisson integral representation implies that for each $\delta \in \mathcal{B}(\mathbb{R})$, $\mu(\delta; ., .)$ is a bilinear form; hence, by (1) and the reflexivity of X, there exists a unique linear transformation

$$E(\delta) : Z \to X$$

such that

$$\mu(\delta; x, x^*) = x^* E(\delta) x, \qquad (x \in Z, x^* \in X^*) \tag{3}$$

and

$$\|E(\delta) x\| \le \|x\|_A \qquad (x \in Z). \tag{4}$$

It follows from (3) and Pettis' theorem that $E(.)x$ is a regular countably additive X-valued measure on $\mathcal{B}(\mathbb{R})$, and we may rewrite (2) in the form

$$[R(t - is; A) - R(t + is; A)]x = \int_{\mathbb{R}} \left[\frac{1}{t - is - u} - \frac{1}{t + is - u}\right] E(du) x \tag{5}$$

for all $t \in \mathbb{R}, s > 0$, and $x \in Z$.

For $x \in Z$ fixed, consider the function

$$F(\zeta) x := \int_{\mathbb{R}} \frac{1}{\zeta - u} E(du) x \qquad (\zeta \in \mathbb{C} - \mathbb{R}).$$

By (4), $F(.)x$ is well-defined, analytic in $\mathbb{C} - \mathbb{R}$, and it follows from (1) that

$$\|F(\zeta) x\| \le \frac{\|x\|_A}{|\Im \zeta|}.$$

In particular, $F(t + is)x \to 0$ as $|s| \to \infty$. By (5),

$$F(t - is)x - F(t + is)x = R(t - is; A)x - R(t + is; A)x \tag{6}$$

for all $t \in \mathbb{R}, s > 0$, and $x \in Z$.

Set $G(.)x = F(.)x - R(.; A)x$. This is an analytic function in $\mathbb{C} - \mathbb{R}$, satisfying $G(\zeta)x = G(\zeta^-)x$ in its domain. Therefore, for each $x^* \in X^*$, the functions $x^* G(\zeta)x$ and $[x^* G(\zeta)x]^- (= [x^* G(\zeta^-)x]^-)$ are both analytic in $\mathbb{C} - \mathbb{R}$. Hence $x^* G(.)x$ is constant there, and since it vanishes as $|\Im \zeta| \to \infty$ (by (1) in Definition 2.9 , and by our previous observation about $F(.)x$), it follows that $G(.)x = 0$ for all $x \in Z$, i.e.,

$$R(\zeta; A) x = \int_{\mathbb{R}} \frac{1}{\zeta - u} E(du) x \tag{7}$$

for all $\zeta \in \mathbb{C} - \mathbb{R}$ and $x \in Z$.

We shall verify now that (7) (i.e., Statement 3. of our theorem) imply all the other statements of Theorem 2.11.

Let $U \in B(X)$ commute with A. Then U commutes with $R(\zeta; A)$ for all non-real ζ. If $x \in Z$, then $R(t + iu; A)Ux = UR(t + iu; A)x \to 0$ as $|u| \to \infty$, for all $t \in \mathbb{R}$. Also, with notations as in Lemma 2.10, we have for fixed $x \in Z$,

$$||x^* P(., s)Ux||_1 = ||x^* UP(., s)x||_1 = ||(U^* x^*)P(., s)x||_1$$

$$= ||V_s U^* x^*||_1 \leq ||V_s|| ||U^* x^*|| \leq ||V_s||.||U||.||x^*||$$

$$\leq ||U||.||x||_A.||x^*||,$$

so that

$$||Ux||_A \leq ||U||.||x||_A \qquad (x \in Z). \tag{8}$$

Thus Z is U-invariant, and by (7),

$$\int_{\mathbb{R}} \frac{1}{\zeta - u} E(du)Ux = R(\zeta; A)Ux = UR(\zeta; A)x$$

$$= \int_{\mathbb{R}} \frac{1}{\zeta - u} UE(du)x$$

for all non-real ζ and $x \in Z$.

By the uniqueness property of the Stieltjes transform (cf. [W]), it follows that $E(\delta)Ux = UE(\delta)x$ for all $x \in Z$ and $\delta \in \mathcal{B}(\mathbb{R})$ (which proves Statement (1) of the theorem).

In particular, taking $U = R(\lambda; A)$ for $\lambda \in \rho(A)$, we obtain

$$R(\lambda; A)E(\mathbb{R})x = E(\mathbb{R})R(\lambda; A)x$$

$$= \lim_{\Im \zeta \to \infty} \int_{\mathbb{R}} \frac{\zeta}{\zeta - u} E(du)R(\lambda; A)x$$

$$= \lim \zeta R(\zeta; A)R(\lambda; A)x = \lim \frac{\zeta}{\zeta - \lambda}[R(\lambda; A) - R(\zeta; A)]x = R(\lambda; A)x,$$

since $x \in Z$ (cf. Condition (1)). Hence

$$E(\mathbb{R})x = x \qquad (x \in Z). \tag{9}$$

By (4), this shows in particular that

$$||x|| \leq ||x||_A \qquad (x \in Z). \tag{10}$$

Therefore, if $\{x_n\} \subset Z$ is $||.||_A$-Cauchy, it is also $||.||$-Cauchy (and of course, $||.||_A$-bounded, say by the constant K). Let $x = \lim_n x_n$ in X. Then by (7) and (1),

$$||R(t + iu; A)x_n|| \leq \frac{||x_n||_A}{|u|} \leq \frac{K}{|u|},$$

and therefore $||R(t + iu; A)x|| \leq \frac{K}{|u|}$. Thus x satisfies Condition (1) in Definition 2.9.

For each $x^* \in X^*$ and $s > 0$, $x^* P(., s)x_n \to_n x^* P(., s)x$ pointwise, so that by Fatou's lemma,

$$||x^* P(., s)x||_1 \leq \liminf_n ||x^* P(., s)x_n||_1$$

$$\leq \liminf_n ||x_n||_A . ||x^*|| \leq K||x^*||,$$

and we conclude that $x \in Z$.

Also, given $\epsilon > 0$, let n_o be such that $||x_n - x_m||_A < \epsilon$ for all $n, m > n_o$. Then $||x^* P(., s)(x_n - x_m)||_1 < \epsilon$ for all unit vectors $x^* \in X^*, s > 0$, and $n, m > n_o$. Letting $m \to \infty$, Fatou's lemma implies that $||x_n - x||_A \leq \epsilon$ for all $n > n_o$, i.e., $x_n \to x$ in the $||.||_A$-norm. We then conclude that $(Z, ||.||_A)$ is a Banach subspace of X.

For $x \in Z$ and $\lambda \in \rho(A)$,

$$\int_{\mathbb{R}} \frac{1}{\zeta - u} E(du) R(\lambda; A)x = R(\zeta; A)R(\lambda; A)x$$

$$= \frac{1}{\zeta - \lambda}[R(\lambda; A) - R(\zeta; A)]x = \frac{1}{\zeta - \lambda} \int_{\mathbb{R}} [\frac{1}{\lambda - u} - \frac{1}{\zeta - u}] E(du)x$$

$$= \int_{\mathbb{R}} \frac{1}{\zeta - u}[\frac{1}{\lambda - u} E(du)x],$$

for all non-real ζ. By the uniqueness property of the Stieltjes transform,

$$E(\delta)R(\lambda; A)x = \int_{\mathbb{R}} \frac{1}{\lambda - u} \chi_\delta(u) E(du)x, \tag{11}$$

for all $x \in Z, \lambda \in \rho(A)$, and $\delta \in \mathcal{B}(\mathbb{R})$ (where χ_δ denotes the characteristic function of δ).

Since $R(\lambda; A)$ commutes with $E(\delta)$, it follows from (11) that for $x \in Z$,

$$R(\lambda; A)E(\delta)x = \int_\delta \frac{1}{\lambda - u} E(du)x \to 0$$

when $|\Im \lambda| \to \infty$.

Also for all unit vectors $x^* \in X^*$, we have by (11)

$$||x^* P(.,s)E(\delta)x||_1 = ||x^* E(\delta)P(.,s)x||_1$$

$$= \frac{s}{\pi} \int_\mathbb{R} |x^* \int_\delta \frac{1}{(t-u)^2 + s^2} E(du)x| dt$$

$$\leq \int_\delta [\frac{s}{\pi} \int_\mathbb{R} \frac{dt}{(t-u)^2 + s^2} |x^* Ex|(du) = |x^* Ex|(\delta) \leq ||\mu(.;x,x^*)|| \leq ||x||_A.$$

Hence $E(\delta)Z \subset Z$ and

$$||E(\delta)x||_A \leq ||x||_A \qquad (x \in Z, \delta \in \mathcal{B}(\mathbb{R})). \tag{12}$$

This shows that $E(\delta) \in T(Z)$; in fact $E(\delta) \in B(Z, ||.||_A)$, with operator norm ≤ 1. By (11) and (7), since $E(\delta)x \in Z$ for $x \in Z$,

$$\int_\mathbb{R} \frac{1}{\lambda-u} \chi_\delta(u) E(du)x = R(\lambda; A)E(\delta)x = \int_\mathbb{R} \frac{1}{\lambda-u} E(du)E(\delta)x.$$

The uniqueness property of the Stieltjes transform implies that

$$E(\sigma)E(\delta)x = \int_\mathbb{R} \chi_\sigma(u)\chi_\delta(u)E(du)x = E(\sigma \cap \delta)x$$

for all $\sigma, \delta \in \mathcal{B}(\mathbb{R})$ and $x \in Z$. We have thus shown that E is a spectral measure on Z.

We prove now Statement 2. in the theorem. The argument yielding (4) in the proof of Theorem 1.49 shows that

$$D(A_Z) = R(\lambda; A)Z \tag{13}$$

for any $\lambda \in \mathbb{C} - \mathbb{R}$.

Let $x \in D(A_Z)$. Write then $x = R(\lambda; A)y$ for a fixed non-real λ and a suitable $y \in Z$. For $-\infty < a < b < \infty$, we have by (11)

$$\int_a^b uE(du)x = \int_a^b uE(du)R(\lambda; A)y = \int_a^b \frac{u}{\lambda-u} E(du)y$$

$$\rightarrow \int_\mathbb{R} \frac{u}{\lambda-u} E(du)y$$

as $a \rightarrow -\infty$ and $b \rightarrow \infty$. Thus $\int_\mathbb{R} uE(du)x$ exists. Writing $\frac{u}{\lambda-u} = \frac{\lambda}{\lambda-u} - 1$, the last relation shows that

$$\int_\mathbb{R} uE(du)x = \lambda \int_\mathbb{R} \frac{1}{\lambda-u} E(du)y - E(\mathbb{R})y$$

$$= \lambda R(\lambda; A)y - y = AR(\lambda; A)y = Ax \in Z \qquad (14)$$

(since $x \in D(A_Z)$). Thus $D(A_Z) \subset Z_1$, where Z_1 denotes the set on the right of Statement 2. of the theorem. On the other hand, if $x \in Z_1$, denote $z = \int_{\mathbb{R}} uE(du)x$; we have $z \in Z$, and for non-real λ, we obtain from (11)

$$R(\lambda; A)z = \lim_{(a,b)} \int_a^b uR(\lambda; A)E(du)x$$

$$= \lim_{(a,b)} \int_a^b \frac{u}{\lambda - u} E(du)x = \int_{\mathbb{R}} \frac{u}{\lambda - u} E(du)x$$

$$= \lambda \int_{\mathbb{R}} \frac{1}{\lambda - u} E(du)x - x = \lambda R(\lambda; A)x - x.$$

Hence
$$x = R(\lambda; A)[\lambda x - z] \in R(\lambda; A)Z = D(A_Z),$$

so that $D(A_Z) = Z_1$. By (14), $Ax = \int_{\mathbb{R}} uE(du)x$ for all $x \in D(A_Z)$.

Finally, let W and F be as in the statement of the theorem. For $x \in W$, the representation 3. (with F) of the resolvent implies (1) in Definition 2.9. Also, for all $x^* \in X^*$,

$$||x^* P(., s)x||_1 = \frac{s}{\pi} \int_{\mathbb{R}} |x^* \int_{\mathbb{R}} \frac{1}{(t-u)^2 + s^2} F(du)x| dt$$

$$\leq \int \frac{s}{\pi} \int \frac{dt}{(t-u)^2 + s^2} |x^* Fx|(du) = |x^* Fx|(\mathbb{R}),$$

so that (2) in Definition 2.9 is satisfied as well, i.e., $x \in Z$. Thus $W \subset Z$, and the uniqueness property of the Stieltjes transform implies that $F(.)x = E(.)x$ for $x \in W.||||$

The discussion preceding Corollary 1.50 yields the following

2.12. COROLLARY. Let A be an operator with real spectrum, acting in the reflexive Banach space X, and let Z be its semi-simplicity manifold. Then $Z = X$ if and only if A is a scalar-type spectral operator. When this is the case, E is the resolution of the identity for A.

We consider now the **operational calculus** τ induced by $E(.)$, the spectral measure on Z, as defined in Section 1.48. In the following, the space Z is normed by $||.||_A$.

2.13. THEOREM. The operational calculus τ is a norm-decreasing algebra homomorphism of $\mathbb{B}(\mathbb{R})$ into $B(Z)$. Moreover, for each $h \in \mathbb{B}(\mathbb{R})$, $\tau(h)$ maps $D(A_Z)$ into itself, and

$$A\tau(h)x = \tau(h)Ax = \int_{\mathbb{R}} uh(u)E(du)x \qquad (x \in D(A_Z)). \qquad (1)$$

Note that the "improper" integral appearing in (1) is $\tau(uh(u))$, defined as usual for functions in $\mathbb{B}_{loc}(\mathbb{R})$.

PROOF. For $x \in Z$ and $h \in \mathbb{B}(\mathbb{R})$, we have by (11) (in the proof of Theorem 2.11)

$$R(\lambda; A)\tau(h)x = \int_{\mathbb{R}} h(u)R(\lambda; A)E(du)x = \int_{\mathbb{R}} \frac{h(u)}{\lambda - u} E(du)x. \qquad (2)$$

Therefore

$$||R(\lambda; A)\tau(h)x|| \leq \frac{||h||_\infty ||x||_A}{|\Im\lambda|} \qquad (\lambda \in \mathbb{C} - \mathbb{R}).$$

In particular, $\tau(h)x$ satisfies 2.9 (1).

By (2), for all $x^* \in X^*$,

$$||x^* P(.,s)\tau(h)x||_1 = ||x^* \int_{\mathbb{R}} p(. - u, s)h(u)E(du)x||_1$$

$$\leq \int\int |h(u)|p(t - u, s)|x^* Ex|(du)dt \leq ||h||_\infty \int\int p(t - u, s)dt|x^* Ex|(du)$$

$$= ||h||_\infty |x^* Ex|(\mathbb{R}) \leq ||h||_\infty ||x||_A ||x^*||,$$

where $|x^* Ex|$ denotes the variation measure of $x^* Ex = \mu(.; x, x^*)$. Thus $\tau(h)x$ satisfies 2.9 (2) as well, i.e., it belongs to Z for all $x \in Z$, and moreover

$$||\tau(h)x||_A \leq ||h||_\infty ||x||_A \qquad (h \in \mathbb{B}(\mathbb{R}), x \in Z). \qquad (3)$$

This establishes that τ is norm-decreasing from $\mathbb{B}(\mathbb{R})$ into $B(Z)$. Since E is a spectral measure on Z, it follows that τ is multiplicative on the simple Borel functions, hence on $\mathbb{B}(\mathbb{R})$ as well.

Next, let $x \in D(A_Z)$. Then $x = R(\lambda; A)y$ for λ non-real and $y \in Z$. Therefore, for any $h \in \mathbb{B}(\mathbb{R})$,

$$\tau(h)x = R(\lambda; A)\tau(h)y \in R(\lambda; A)Z = D(A_Z),$$

i.e., $\tau(h)D(A_Z) \subset D(A_Z)$.

In particular, for h, x, y as before, by multiplicativity of τ on $\mathbb{B}(\mathbb{R})$, the limit

$$\lim_{(a,b)} \int_a^b u E(du) \tau(h) x = \lim_{(a,b)} \int_a^b u h(u) E(du) x$$

exists in X and belongs to Z, and equals $A\tau(h)x$ (by Theorem 2.11). Finally, we observe that the bounded operator $AR(\lambda; A) = \lambda R(\lambda; A) - I$ commutes with E, hence with $\tau(h)$, and therefore

$$A\tau(h)x = AR(\lambda; A)\tau(h)y = \tau(h)AR(\lambda; A)y = \tau(h)Ax.||||$$

Taking in particular the functions $h_t(u) = e^{itu}$ $(t, u \in \mathbb{R})$, let

$$T(t) = \tau(h_t) \qquad (t \in \mathbb{R}).$$

Then $T(.)$ is a group of contractions in Z, which map $D(A_Z)$ into itself. It is continuous with respect to the X-norm (as follows at once by dominated convergence), but not necessarily with respect to the Z-norm.

Consider the continuous functions on $\mathbb{R}$

$$k_t(u) = \frac{e^{itu} - 1}{itu} \qquad (t, u \neq 0),$$

and $k_t(0) = 1$ (for $t \neq 0$). We have $|k_t(u)| \leq 1$ and $k_t(u) \to 1$ as $t \to 0$ (for all $u \in \mathbb{R}$). For $x \in D(A_Z)$, we apply (1) (in Theorem 2.13) and the Dominated Convergence Theorem for vector measures:

$$t^{-1}[T(t)x - x] = i \int_{\mathbb{R}} u k_t(u) E(du) x = i\tau(k_t) Ax$$

$$= i \int_{\mathbb{R}} k_t(u) E(du) Ax \to_{t \to 0} iAx.$$

(limit in X). Also

$$s^{-1}[T(t+s)x - T(t)x] = s^{-1}[T(s)T(t)x - T(t)x] \to_{s \to 0} iAT(t)x$$

(limit in X), since $T(t)x \in D(A_Z)$ for $x \in D(A_Z)$, by Theorem 2.13. We formalize the above discussion in

2.14. COROLLARY. $T(.)$ is a group of contractions in Z, continuous in the X-topology on Z, and leaving $D(A_Z)$ invariant. Moreover, in that topology, the

91

generator of $T(.)$ coincides with iA on $D(A_Z)$, and $u := T(.)x$ solves the ACP on $\mathbb{R}$

$$u' = iAu, \qquad u(0) = x$$

for $x \in D(A_Z)$.

The basic properties of the operational calculus τ on $\mathbb{B}_{loc}(\mathbb{R})$ are collected in the following

2.15. THEOREM. (i) $\tau(\lambda h) = \lambda \tau(h)$ $\qquad$ $(0 \neq \lambda \in \mathbb{C}, h \in \mathbb{B}_{loc}(\mathbb{R}))$;
(ii) $D(\tau(h) + \tau(g)) = D(\tau(h + g)) \cap D(\tau(g))$, and

$$\tau(h + g)x = \tau(h)x + \tau(g)x$$

for all $x \in D(\tau(h) + \tau(g))$ and $h, g \in \mathbb{B}_{loc}(\mathbb{R})$;
(iii) $E(\delta)D(\tau(h)) \subset D(\tau(h))$, and

$$\tau(h)E(\delta)x = E(\delta)\tau(h)x = \tau(h\chi_\delta)x \qquad (x \in D(\tau(h))),$$

for all compact $\delta \subset \mathbb{R}$ and $h \in \mathbb{B}_{loc}(\mathbb{R})$;
(iv) $D(\tau(h)\tau(g)) = D(\tau(hg)) \cap D(\tau(g))$, and

$$\tau(h)\tau(g)x = \tau(hg)x$$

for all $x \in D(\tau(h)\tau(g))$ and $h, g \in \mathbb{B}_{loc}(\mathbb{R})$.

PROOF. (i) is trivial.
In the following, h, g will denote arbitrary functions in $\mathbb{B}_{loc}(\mathbb{R})$.
Proof of (ii). Let $x \in D(\tau(h) + \tau(g)) := D(\tau(h)) \cap D(\tau(g))$. Then

$$\lim_{a,b} \int_a^b h(u)E(du)x$$

exists in X and belongs to Z, and similarly for g. Therefore $\lim_{a,b}$ of the sum of the two integrals (i.e., of $\int_a^b (h + g)(u)E(du)x$) exists and belongs to Z. Thus $x \in D(\tau(h+g))$ and $\tau(h+g)x = \tau(h)x + \tau(g)x$. On the other hand, if $x \in D(\tau(h+g)) \cap D(\tau(g))$, then writing $\int_a^b h(u)E(du)x = \int_a^b (h+g)(u)E(du)x - \int_a^b g(u)E(du)x$, we see that $x \in D(\tau(h))$, so we have the wanted equality of domains.

(iii) Let $\delta \subset \mathbb{R}$ be compact, and $x \in D(\tau(h))$. In particular, $x \in Z$, and therefore $E(\delta)x \in Z$. Since τ is multiplicative on $\mathbb{B}(\mathbb{R})$, we have

$$\int_a^b h(u)E(du)E(\delta)x = \tau(h\chi_{[a,b]})\tau(\chi_\delta)x = \tau(h\chi_{[a,b]\cap\delta})$$

92

$$= \int_{\mathbb{R}} h(u)\chi_{[a,b]\cap\delta}(u)E(du)x.$$

In the last integral, the integrand is majorized by the bounded function $|h|$, and converges pointwise to $h\chi_\delta$ when $a \to -\infty$ and $b \to \infty$. By dominated convergence for vector measures, it follows that $\lim_{a,b}$ of that integral exists in X and equals $\tau(h\chi_\delta)x \in Z$ (since $h\chi_\delta \in \mathbb{B}(\mathbb{R})$). Hence $E(\delta)x \in D(\tau(h))$ and $\tau(h)E(\delta)x = \tau(h\chi_\delta)x$.

We also have $\tau(h)x \in Z$, because $x \in D(\tau(h))$. Therefore, by (11) in the proof of Theorem 2.11, we have for $\lambda \in \mathbb{C} - \mathbb{R}$

$$\int_{\mathbb{R}} \frac{1}{\lambda - u} E(du)\tau(h)x = R(\lambda; A)\tau(h)x$$

$$= R(\lambda; A)\lim_{a,b} \int_a^b h(u)E(du)x = \lim_{a,b} \int_a^b h(u)E(du)R(\lambda; A)x$$

$$= \lim_{a,b} \int_a^b \frac{h(u)}{\lambda - u} E(du)x = \int_{\mathbb{R}} \frac{h(u)}{\lambda - u} E(du)x.$$

Hence

$$E(\delta)\tau(h)x = \int_{\mathbb{R}} h(u)\chi_\delta(u)E(du)x = \tau(h\chi_\delta)x,$$

by the uniqueness property of the Stieltjes transform. This completes the proof of (iii).

(iv) Let $x \in D(\tau(h)\tau(g))$, i.e., $x \in D(\tau(g))$ and $\tau(g)x \in D(\tau(h))$. By the multiplicativity of τ on $\mathbb{B}(\mathbb{R})$ and by (iii),

$$\int_a^b h(u)g(u)E(du)x = \tau(h\chi_{[a,b]}\cdot g\chi_{[a,b]})x$$

$$= \tau(h\chi_{[a,b]})\tau(g\chi_{[a,b]})x = \tau(h\chi_{[a,b]})E([a,b])\tau(g)x$$

$$= \tau(h\chi_{[a,b]})\tau(g)x \to \tau(h)\tau(g)x$$

when $a \to -\infty$ and $b \to \infty$, since $\tau(g)x \in D(\tau(h))$. Hence $x \in D(\tau(hg))$ and $\tau(hg)x = \tau(h)\tau(g)x$.

In particular, $D(\tau(h)\tau(g)) \subset D(\tau(hg)) \cap D(\tau(g))$.

On the other hand, if x belongs to the right side of the last relation, we have in particular $\tau(g)x \in Z$. Therefore, by the multiplicativity of τ on $\mathbb{B}(\mathbb{R})$ and by (iii), we have

$$\tau(h\chi_{[a,b]})\tau(g)x = \tau(h\chi_{[a,b]}^2)\tau(g)x$$

$$= \tau(h\chi_{[a,b]})E([a,b])\tau(g)x = \tau(h\chi_{[a,b]})\tau(g\chi_{[a,b]})x$$

93

$$= \tau(hg\chi_{[a,b]})x \to_{a,b} \tau(hg)x \in Z,$$

because $x \in D(\tau(hg))$. Hence $\tau(g)x \in D(\tau(h))$.||||

We show next that τ operates in the desired way on polynomials. If $p(u) = \Sigma_o^n \alpha_k u^k$ with $n \geq 1$ and $\alpha_n \neq 0$, we define as usual

$$p(A_Z) := \Sigma_o^n \alpha_k A_Z^k = \Sigma_o^n \alpha_k A^k$$

restricted to

$$D(p(A_Z)) = D(A_Z^n) := \{x \in D(A_Z^{n-1}); A_Z^{n-1}x \in D(A_Z)\}.$$

2.16. THEOREM. 1. $D(p(A_Z)) = \bigcap_{k=1}^n D(\tau(u^k))$;
2. $p(A_Z)x = \tau(p)x$ for all $x \in D(p(A_Z))$.

PROOF. We first prove the following

LEMMA. For $n = 1, 2, \ldots$ and any $\lambda \in \rho(A)$,
(i) $D(A_Z^n) = \{x \in D(A^n); A^k x \in Z, k = 0, 1, \ldots, n\}$;
(ii) $D(A_Z^n) = R(\lambda; A)^n Z$.

PROOF (of Lemma). (i) is easily verified by induction. The validity of (ii) for $n = 1$ was observed before. Assume (ii) for $n - 1$ (where $n \geq 2$). Since Z is $R(\lambda; A)$-invariant,

$$R(\lambda; A)^n Z \subset R(\lambda; A)^{n-1} Z = D(A_Z^{n-1}),$$

by the induction hypothesis.
Let $x \in R(\lambda; A)^n Z$; then $x \in D(A_Z^{n-1})$, and writing $x = R(\lambda; A)^n y$ with $y \in Z$, we have

$$A^{n-1}x = [AR(\lambda; A)]^{n-1} R(\lambda; A)y = [\lambda R(\lambda; A) - I]^{n-1} R(\lambda; A)y$$

$$= R(\lambda; A)[\lambda R(\lambda; A) - I]^{n-1}y \in D(A) \cap Z.$$

Hence $x \in D(A^n)$ and

$$A^n x = [AR(\lambda; A)]^n y = [\lambda R(\lambda; A) - I]^n y \in Z.$$

By (i), this shows that $x \in D(A_Z^n)$.
On the other hand, if $x \in D(A_Z^n)$, then by (i), $x \in D(A^n)$ and $A^k x \in Z$ for $k = 0, \ldots, n$. Therefore $y := (\lambda I - A)^n x \in Z$, and $x = R(\lambda; A)^n y \in R(\lambda; A)^n Z$ (for $\lambda \in \rho(A)$), and (ii) follows for n.||||

94

Back to the proof of the theorem, let $x \in D(A_Z^n)$ and fix $\lambda \in \rho(A)$. By the lemma, write $x = R(\lambda; A)^n y$ with $y \in Z$. Applying (11) in the proof of Theorem 2.11 repeatedly, we obtain

$$E(du)x = E(du)R(\lambda; A)^k[R(\lambda; A)^{n-k}y] = (\lambda - u)^{-k}E(du)[R(\lambda; A)^{n-k}y]$$

for $k = 1, ..., n$, since $R(\lambda; A)^{n-k}y \in Z$.

Hence, for $-\infty < a < b < \infty$,

$$\int_a^b u^k E(du)x = \int_a^b (\frac{u}{\lambda - u})^k E(du)[R(\lambda; A)^{n-k}y]$$

$$\to_{a,b} \int_{\mathbb{R}} (\frac{u}{\lambda - u})^k E(du)[R(\lambda; A)^{n-k}y], \tag{1}$$

since $[u/(\lambda - u)]^k$ is a bounded function of u on $\mathbb{R}$.

Thus $\int_{\mathbb{R}} u^k E(du)x$ exists (in X) and equals the integral in (1), which belongs to Z for $k = 1, ..., n$, by Theorem 2.13. This proves the inclusion $\subset$ in Statement 1. of the theorem.

Next, let x belong to the set on the right of 1. For each $k = 0, ..., n$, denote

$$z_k = \int_{\mathbb{R}} u^k E(du)x \quad (\in Z).$$

By (11) in the proof of Theorem 2.11, we have for $k = 1, ..., n$,

$R(\lambda; A)z_k = \lim_{a,b} \int_a^b u^k R(\lambda; A)E(du)x$
$= \lim_{a,b} \int_a^b u^{k-1} \frac{u}{\lambda - u} E(du)x$
$= \lim_{a,b} \int_a^b u^{k-1}(\frac{\lambda}{\lambda - u} - 1)E(du)x$
$= \lim_{a,b}[\lambda R(\lambda; A) - I]\int_a^b u^{k-1}E(du)x$
$= [\lambda R(\lambda; A) - I]z_{k-1}.$

Therefore

$$z_{k-1} = R(\lambda; A)(\lambda z_{k-1} - z_k) \in D(A) \cap Z,$$

and

$$Az_{k-1} = [\lambda R(\lambda; A) - I](\lambda z_{k-1} - z_k) = \lambda R(\lambda; A)z_k - [\lambda R(\lambda; A) - I]z_k = z_k,$$

for $k = 1, ..., n$.

Since $z_0 = x$, it follows from the above recursion that $x \in D(A^n)$ and $A^k x = z_k \in Z$ for $k \leq n$, i.e., $x \in D(A_Z^n)$ by Part (i) of the lemma. This proves Statement 1. of the theorem, and also the relation

$$A^k x = \int_{\mathbb{R}} u^k E(du)x \quad (x \in D(A_Z^k), \quad k = 1, 2, ...),$$

which clearly implies Statement 2.||||

C. SEMI-SIMPLICITY MANIFOLD (Case $\mathbb{R}^+ \subset \rho(-A)$)

We generalize the construction of the semi-simplicity manifold to operators $-A$ with spectrum in a half-plane, say in the closed left half-plane, to fix the ideas. Actually, all we need for our construction is that $\mathbb{R}^+ := (0, \infty)$ be contained in the resolvent set of $-A$. Let then

$$R(t) := R(t; -A) \qquad (t > 0),$$

and

$$S := AR(I - AR).$$

The function $S(t) = tR(t)[I - tR(t)]$ is a well-defined $B(X)$-valued function on $\mathbb{R}^+$, and for all $k = 1, 2, ...$, the powers S^k are of class C^∞.

In the following discussion, the $L^1(\mathbb{R}^+, \frac{dt}{t})$- norm is denoted by $||.||_1$.
The Beta function is

$$B(s, t) := \frac{\Gamma(s)\Gamma(t)}{\Gamma(s + t)} \qquad (s, t \in \mathbb{R}^+).$$

2.17. DEFINITION. Let $-A$ be an operator with $(0, \infty) \subset \rho(-A)$, and let S be the operator function defined above. The "semi-simplicity manifold" for $-A$ is the set Z of all $x \in X$ such that

$$\sup_{k \in \mathbb{N}} \frac{||x^* S^k x||_1}{B(k, k)} < \infty$$

for all $x^* \in X^*$.

Using the Closed Graph Theorem, Fatou's lemma, and the Uniform Boundedness Theorem as in the proof of Lemma 2.10, we obtain

2.18. LEMMA. For all $x \in Z$,

$$||x||_Z := \sup\{\frac{||x^* S^k x||_1}{B(k, k)}, ||x||; k \in \mathbb{N}, ||x^*|| = 1\} < \infty.$$

2.19. LEMMA. The space $Z := (Z, ||.||_Z)$ is a Banach subspace of X, invariant for any $U \in B(X)$ commuting with A, and $||U||_{B(Z)} \leq ||U||_{B(X)}$.

PROOF. The proof is analogous to the one we gave for the real-spectrum case (see proof of Theorem 2.11).

2.20. THEOREM. Let $-A$ be an operator in the reflexive Banach space X, whose resolvent set contains the axis $\mathbb{R}^+$, and let Z be its semi-simplicity manifold. Then there exists a spectral measure on Z,

$$E(.) : \mathcal{B}(\mathbb{R}^+) \to \mathcal{T}(Z),$$

such that
  1. for each $\delta \in \mathcal{B}(\mathbb{R}^+)$, $E(\delta)$ commutes with every $U \in B(X)$ which commutes with A;
  2. (i) $D(A_Z) = \{x \in Z ; \lim_{b\to\infty} \int_0^b sE(ds)x$ exists in X and belongs to $Z\}$, and
  (ii) $Ax = \int_0^\infty sE(ds)x \qquad (x \in D(A_Z))$,
  where the last integral is defined as the limit in (i);
  3. $R(t)x = \int_0^\infty \frac{1}{t+s}E(ds)x \qquad (x \in Z, t > 0)$.
  Moreover, Z is *maximal-unique* relative to Property 3., in the sense of Theorem 2.11.

PROOF. Let L_k be the Widder formal differential operators

$$L_k := c_k M^{k-1} D^{2k-1} M^k \qquad (k \in \mathbb{N}),$$

where

$$M : f(t) \to tf(t); \qquad D : f \to f'$$

are respectively the "multiplication" and the differentiation operators acting on functions of $t \in \mathbb{R}^+$. The constants c_k are given by $c_1 = 1$ and

$$c_k = \frac{(-1)^{k-1}}{\Gamma(k-1)\Gamma(k+1)} \qquad (k \geq 2).$$

By Leibnitz' rule,

$$L_k = c_k' \Sigma_{j=0}^k \Gamma(k+j)^{-1} \binom{k}{j} M^{k+j-1} D^{k+j-1},$$

where $c_1' = 1$ and $c_k' = (-1)^{k-1} B(k-1, k+1)^{-1}$ for $k \geq 2$. Since

$$D^{k+j-1}(x^* Rx) = (-1)^{k+j-1}\Gamma(k+j)x^* R^{k+j}x,$$

we have

$$L_k x^* R(t)x = c_k'' t^{-1} x^* (tR)^k \Sigma_{j=0}^k \binom{k}{j} (-tR)^j x = c_k'' t^{-1} x^* S^k(t)x,$$

where $c_1'' = 1$ and $c_k'' = B(k-1, k+1)^{-1}$ for $k \geq 2$. Therefore, for $x \in Z$ and $x^* \in X^*$,

$$\int_0^\infty |L_k(x^* Rx)|dt = c_k'' ||x^* S^k x||_1 \leq ||x||_Z ||x^*||, \tag{1}$$

trivially for $k = 1$, and because $\frac{B(k,k)}{B(k-1,k+1)} = \frac{k-1}{k} < 1$ for $k > 1$. We now rely on the following complex version of Widder's theorem (cf. [W; Theorem 16, p. 361]): Let f be a C^∞ complex function on $\mathbb{R}^+$, such that

$$K := \sup_{k \in \mathbb{N}} \int_0^\infty |L_k f| dt < \infty.$$

Then the limit $c = \lim_{t \to 0+} tf(t)$ exists, and there exists a unique complex regular Borel measure μ on $\mathbb{R}^+$ such that $||\mu|| \leq 2K + |c|$ and

$$f(t) = \int_0^\infty \frac{\mu(ds)}{t+s} \qquad (t \in \mathbb{R}^+).$$

Taking $f = x^* Rx$ with $x \in Z$ and $x^* \in X^*$ fixed, we have $K \leq ||x||_Z ||x^*|| < \infty$ by (1). Thus $\lim_{t \to 0+} x^* tR(t)x$ exists for each $x^* \in X^*$. By the Uniform Boundedness Theorem,

$$H_x := \sup_{0 < t \leq 1} ||tR(t)x|| < \infty \qquad (x \in Z). \tag{2}$$

Let $M_x := 2||x||_Z + H_x$. By Widder's theorem, there exists a unique complex regular Borel measure $\mu(.; x, x^*)$ such that

$$||\mu(.; x, x^*)|| \leq M_x ||x^*||$$

and

$$x^* R(t)x = \int_0^\infty \frac{\mu(ds; x, x^*)}{t+s} \qquad (t \in \mathbb{R}^+),$$

for all $x \in Z$ and $x^* \in X^*$.

This implies in particular that

$$||tR(t)x|| \leq M_x \qquad (t \in \mathbb{R}^+). \tag{2'}$$

The uniqueness of the Stieltjes transform implies that for each fixed $\delta \in \mathcal{B}(\mathbb{R}^+)$ and $x \in Z$, $\mu(\delta; x, .)$ is a continuous linear functional on X^*, so that, by reflexivity of

98

X, there exists a unique function $E(.)x : \mathcal{B}(\mathbb{R}^+) \to X$ (for each fixed $x \in Z$) such that

$$\mu(.;x,x^*) = x^*E(.)x \qquad (x^* \in X^*).$$

Necessarily, $E(\delta)$ is a linear operator with domain Z, and

$$\|E(\delta)x\| \le M_x \qquad (\delta \in \mathcal{B}(\mathbb{R}^+), x \in Z).$$

By Pettis' theorem, $E(.)x$ is a strongly countably additive vector measure, and

$$R(t)x = \int_0^\infty \frac{E(ds)x}{t+s} \qquad (t > 0, x \in Z).$$

This is Property 3., which corresponds to (7) in the proof of Theorem 2.11. As in the latter case, we shall see that it implies that E is a spectral measure on Z satisfying Properties 1. and 2. of our theorem.

Property 1. is an immediate consequence of the uniqueness property of the Stieltjes transform. Taking then, in particular, $U = R(u)$ for $u > 0$ fixed, we obtain for $x \in Z$

$$R(u)E(\mathbb{R}^+)x = E(\mathbb{R}^+)R(u)x = \lim_{t \to \infty} \int_0^\infty \frac{t}{t+s}E(ds)R(u)x$$

$$= \lim_t tR(t)R(u)x = \lim_t \frac{t}{t-u}R(u)x - \lim_t \frac{tR(t)x}{t-u} = R(u)x,$$

by the resolvent equation and (2'). Since $R(u)$ is one-to-one, it follows that $E(\mathbb{R}^+) = I/Z$.

For $t, u > 0, t \ne u$, and $x \in Z$, we have by Property 3., the resolvent equation, and the fact that $R(u)x \in Z$,

$$\int_0^\infty \frac{1}{t+s}E(ds)R(u)x = R(t)R(u)x$$

$$= \frac{1}{t-u}\int_0^\infty [\frac{1}{u+s} - \frac{1}{t+s}]E(ds)x = \int_0^\infty \frac{1}{t+s}[\frac{1}{u+s}E(ds)x].$$

By the uniqueness of the Stieltjes transform,

$$E(ds)R(u)x = \frac{1}{u+s}E(ds)x, \qquad (3)$$

and inductively,

$$E(ds)R(u)^k x = \frac{1}{(u+s)^k}E(ds)x,$$

for all $k \in \mathbb{N}, u > 0$, and $x \in Z$. Therefore

$$E(ds)p(R(u))x = p(\frac{1}{u+s})E(ds)x$$

for all polynomials p. In particular,

$$E(ds)S^k(u)x = \frac{u^k}{(u+s)^k}[1 - \frac{u}{u+s}]^k E(ds)x$$

$$= \frac{(us)^k}{(u+s)^{2k}}E(ds)x,$$

for all $u > 0, k \in \mathbb{N}$, and $x \in Z$. Property 1. for $U = S^k(u)$ implies then that

$$x^* S^k(u)E(\delta)x = \int_\delta \frac{(us)^k}{(u+s)^{2k}} x^* E(ds)x.$$

By Tonelli's theorem,

$$||x^* S^k E(\delta)x||_1 \leq \int_\delta \int_0^\infty \frac{(us)^k}{(u+s)^{2k}} \frac{du}{u} |x^* E(.)x|(ds)$$

$$= \int_\delta \int_0^\infty \frac{t^k}{(1+t)^{2k}} \frac{dt}{t} |x^* E(.)x|(ds) = B(k,k)|x^* Ex|(\delta)$$

$$\leq B(k,k)M_x||x^*||,$$

for all $x^* \in X^*, k \in \mathbb{N}, \delta \in \mathcal{B}(\mathbb{R}^+)$, and $x \in Z$. Therefore $E(\delta)x \in Z$ for $x \in Z$ (i.e., $E(\delta) \in T(Z)$), and

$$||E(\delta)x||_Z \leq M_x \qquad (x \in Z, \delta \in \mathcal{B}(\mathbb{R}^+)). \tag{4}$$

It now follows from Property 3. with the vector $E(\delta)x \in Z$ (whenever $x \in Z$), Property 1. (with $U = R(u)$ for any $u > 0$), and (3):

$$R(u)E(\delta)x = \int_0^\infty \frac{1}{u+s} E(ds)E(\delta)x$$

$$= E(\delta)R(u)x = \int_0^\infty \frac{1}{u+s}\chi_\delta(s)E(ds)x,$$

and therefore, by the uniqueness of the Stieltjes transform,

$$E(\sigma)E(\delta)x = \int_0^\infty \chi_\sigma(s)\chi_\delta(s)E(ds)x = E(\sigma \cap \delta)x$$

for all $\sigma, \delta \in \mathcal{B}(\mathbb{R}^+)$. In conclusion, E is a spectral measure on Z.

Since $D(A_Z) = R(t)Z$ for any $t > 0$, write any given $x \in D(A_Z)$ as $x = R(t)y$ for a fixed $t > 0$ and a suitable $y \in Z$. Then

$$\int_0^b sE(ds)x = \int_0^b \frac{s}{t+s}E(ds)y$$

$$\to_{b \to \infty} \int_0^\infty \frac{s}{t+s}E(ds)y$$

$$= \int_0^\infty [1 - \frac{t}{t+s}]E(ds)y$$

$$= [I - tR(t)]y(\in Z)$$

$$= AR(t)y = Ax.$$

If Z_1 denotes the set on the right of Property 2(i), we obtained that $D(A_Z) \subset Z_1$ and Property 2(ii) is valid on $D(A_Z)$. On the other hand, if $x \in Z_1$, denote the limit in Property 2(i) by $z \in Z$. Then for any $t > 0$,

$$R(t)z = \lim_{b \to \infty} \int_0^b sR(t)E(ds)x = \lim_{b \to \infty} \int_0^b \frac{s}{t+s}E(ds)x$$

$$= \int_0^\infty \frac{s}{t+s}E(ds)x = x - tR(t)x.$$

Therefore $x = R(t)[z + tx] \in R(t)Z = D(A_Z)$, so that $D(A_Z) = Z_1$.

Suppose now that W is a linear manifold in X and F is a spectral measure on W with Property 3. of E. Fix $x \in W$. Differentiating repeatedly, we obtain

$$R^k(t)x = \int_0^\infty (\frac{1}{t+s})^k F(ds)x$$

for all $k = 1, 2, \ldots$ and $t > 0$. Therefore

$$p(R(t))x = \int_0^\infty p(\frac{1}{t+s})F(ds)x \qquad (5)$$

for all polynomials p and $t > 0$. In particular,

$$S^k(t)x = \{tR(t)[1 - tR(t)]\}^k = \int_0^\infty \{\frac{t}{t+s}[1 - \frac{t}{t+s}]\}^k F(ds)x$$

$$= \int_0^\infty \frac{(ts)^k}{(t+s)^{2k}}F(ds)x.$$

Therefore, for all $x^* \in X^*$ and $k \in \mathbb{N}$, we have by Tonnelli's theorem

$$\frac{||x^* S^k x||_1}{B(k,k)} \leq B(k,k)^{-1} \int_0^\infty \int_0^\infty \frac{(ts)^k}{(t+s)^{2k}} |x^* Fx|(ds) \frac{dt}{t}$$

$$= B(k,k)^{-1} \int_0^\infty \int_0^\infty \frac{u^k}{(1+u)^{2k}} \frac{du}{u} |x^* Fx|(ds)$$

$$= ||x^* Fx|| < \infty.$$

Hence $x \in Z$, i.e., $W \subset Z$. Also, for all $x \in W \subset Z$, we have

$$R(t)x = \int_0^\infty \frac{1}{t+s} F(ds)x = \int_0^\infty \frac{1}{t+s} E(ds)x \qquad (t > 0),$$

and therefore $F(\delta)x = E(\delta)x$ for all $\delta \in \mathcal{B}(\mathbb{R}^+)$, by the uniqueness property of the Stieltjes transform. $||||$

As before, the important special case $Z = X$ gives the following result.

2.21. THEOREM. Let $-A$ be an operator with $\mathbb{R}^+ \subset \rho(-A)$, acting in the reflexive Banach space X, and let Z be its semi-simplicity manifold. Then the following statements are equivalent:
(a) $Z = X$.
(b) $K := \sup_{||x||=1} ||x||_Z < \infty$.
(c) A is spectral of scalar type, with spectrum in $[0, \infty)$.

PROOF. Since $||.|| \leq ||.||_Z$, the equivalence of (a) and (b) follows from the Closed Graph Theorem.
Assume now (a). By (2) and the Uniform Boundedness Theorem,

$$H := \sup_{0 < t \leq 1} ||tR(t)|| < \infty,$$

and so

$$M_x \leq M||x|| \qquad (x \in X),$$

where $M := 2K + H$.
It now follows from (4) that $E(\delta) \in B(X)$ (actually, $||E(\delta)||_{B(X)} \leq M$) for all $\delta \in \mathcal{B}(\mathbb{R}^+)$. Therefore E is a spectral measure in the usual sense, and Property 2. of Theorem 2.20 just states that A is a scalar-type spectral operator with resolution of the identity E, and then necessarily $\sigma(A) \subset [0, \infty)$ (cf. [DS,III]).
We show finally that (c) implies (a) (even without the reflexivity hypothesis). Let E be the resolution of the identity of the scalar-type operator A with spectrum

in $[0, \infty)$. Then E is a spectral measure on X satisfying Property 3. of the theorem (on X). By the maximality property of Z, we have necessarily $X = Z.\|\|\|$

The operational calculus results contained in Theorems 2.13, 2.15, and 2.16, and in Corollary 2.14 (with the obvious modification), are generalized in a routine way to the present situation (i.e., with the assumption $\mathbb{R}^+ \subset \rho(-A)$).

Observe that Theorem 2.20 applies in particular to the case where $-A$ generates a C_o-semigroup of contractions, $T(.)$. In that case, by (5) (for the spectral measure on Z, $E(.)$), we have

$$[\frac{n}{t}R(\frac{n}{t})]^n x = \int_0^\infty [\frac{n}{t} \frac{1}{\frac{n}{t}+s}]^n E(ds)x = \int_0^\infty \frac{E(ds)x}{[1+\frac{ts}{n}]^n}$$

$$\to_{n\to\infty} \int_0^\infty e^{-ts} E(ds)x$$

for all $x \in Z$ and $t > 0$, by the Lebesgue Dominated Convergence Theorem for vector measures. By Theorem 1.36, it follows that

$$T(t)x = \int_0^\infty e^{-ts} E(ds)x \qquad (x \in Z, t \ge 0),$$

that is, $T(.)x$ is the Laplace-Stieltjes transform of the vector measure $E(.)x$ for all $x \in Z$.

We may consider the more general question of constructing a maximal Banach subspace on which an arbitrary family of closed operators is the Laplace-Stieltjes transform of a vector measure. This will be done in the next section.

D. LAPLACE-STIELTJES SPACE

Denote by $\mathcal{L}$ the Laplace transform,

$$(\mathcal{L}\phi)(t) := \int_0^\infty e^{-ts}\phi(s)ds \qquad (t \geq 0),$$

acting on a space of functions to be specified as we proceed. We may choose for example the space

$$C_c^\infty := C_c^\infty(\mathbb{R}^+)$$

of all complex C^∞-functions with compact support in $\mathbb{R}^+$.

Let $K(X)$ denote the set of all closed operators acting on X.

2.22. DEFINITION. Let $F : [0, \infty) \to K(X)$ be such that $F(0) = I$. The "Laplace-Stieltjes space" for F is the set W of all x in the "common domain" of F,

$$\mathcal{D} := \bigcap_{s>0} D(F(s)),$$

such that $F(.)x$ is strongly continuous on $[0, \infty)$, and

$$\|x\|_W := \sup\{\|\int_0^\infty \phi(s)F(s)x\,ds\|; \phi \in C_c^\infty, \|\mathcal{L}\phi\|_\infty = 1\}$$

is finite.

2.23. THEOREM. Let W be the Laplace-Stieltjes space for F, as in Definition 2.22, normed by $\|.\|_W$. Then W is a Banach subspace of X, and in case X is reflexive, there exits a uniquely determined function E on $\mathcal{B}([0, \infty))$ into the closed unit ball $B(W, X)_1$ of $B(W, X)$, such that
  (i) for each $x \in W$, $E(.)x$ is a regular countably additive X-valued measure, and
  (ii) $F(t)x = \int_0^\infty e^{-ts}E(ds)x$ for all $t \geq 0$ and $x \in W$.
  (iii) If $T \in B(X)$ leaves the common domain $\mathcal{D}$ invariant and commutes with $F(s)|_\mathcal{D}$ for all $s > 0$, then $T \in B(W)$ (with $\|T\|_{B(W)} \leq \|T\|_{B(X)}$) and $TE(\delta) = E(\delta)T$ on W, for all $\delta \in \mathcal{B}([0, \infty))$.
  Moreover, the pair (W, E) is maximal-unique in the following sense: if (Y, E') is a pair with the properties of (W, E) (not including (iii)), then $(Y, E') \subset (W, E)$,

meaning that Y is continuously imbedded in W and $E'(\delta) = E(\delta)|_Y$ for all $\delta \in \mathcal{B}([0, \infty))$.

The proof depends on a general criterion for belonging to the range of the adjoint T^* of a densely defined operator T.

2.24. LEMMA. Let $\mathcal{E}, \mathcal{F}$ be normed spaces, and let $T : \mathcal{E} \to \mathcal{F}$ be a densely defined linear operator. Let $u^* \in \mathcal{E}^*$ and $M > 0$ be given. Then there exists $v^* \in D(T^*)$ with $||v^*|| \leq M$ such that $u^* = T^*v^*$ if and only if

$$|u^*u| \leq M||Tu|| \qquad (u \in D(T)). \qquad (*)$$

PROOF. If $u^* = T^*v^*$ with $v^* \in D(T^*)$ such that $||v^*|| \leq M$, then for all $u \in D(T)$,

$$|u^*u| = |(T^*v^*)(u)| = |v^*(Tu)|$$
$$\leq ||v^*||.||Tu|| \leq M||Tu||.$$

Conversely, if $(*)$ is satisfied, define

$$\pi : ran\,(T) \to \mathbb{C}$$

by

$$\pi(Tu) = u^*u \qquad (u \in D(T)).$$

If $u, u' \in D(T)$ are such that $Tu = Tu'$, then by $(*)$,

$$|u^*u - u^*u'| = |u^*(u - u')| \leq M||T(u - u')|| = 0,$$

so that π is well-defined. It is linear and bounded on $ran(T)$, with norm $\leq M$ (by $(*)$). By the Hahn-Banach Theorem, there exists $v^* \in \mathcal{F}^*$ such that $||v^*|| \leq M$ and

$$v^*|_{ran\,(T)} = \pi.$$

Thus

$$v^*(Tu) = u^*u \qquad (u \in D(T)).$$

This shows that $v^* \in D(T^*)$ and $T^*v^* = u^*$.||||

Note that for $T \in B(\mathcal{E}, \mathcal{F})$, Condition $(*)$ needs to be required only for all u in a *dense* subset of $\mathcal{E}$.

We apply the lemma to the Laplace transform:

2.25. LEMMA. A function $h : [0, \infty)$ is the Laplace-Stieltjes transform $h(t) = \int_0^\infty e^{-ts}\mu(ds)$ of a regular complex Borel measure μ on $[0, \infty)$ with total variation norm $||\mu|| \leq M$ if and only if it is continuous and

$$|\int_0^\infty h(t)\phi(t)dt| \leq M||\mathcal{L}\phi||_\infty$$

for all $\phi \in C_c^\infty(\mathbb{R}^+))$.

PROOF. If h is a Laplace-Stieltjes transform, it is certainly continuous, so that the integrals $\int_0^\infty h(t)\phi(t)dt$ make sense for all $\phi \in C_c^\infty$, and

$$|\int_0^\infty h(t)\phi(t)dt| = |\int_0^\infty (\mathcal{L}\phi)(s)\mu(ds)|$$

$$\leq ||\mu||.||\mathcal{L}\phi||_\infty \leq M||\mathcal{L}\phi||_\infty.$$

For the converse, apply Lemma 2.24 to the operator

$$\mathcal{L} : L^1([0, \infty)) \to C_0([0, \infty),$$

where $C_0([0, \infty))$ denotes the space of all complex continuous functions on $[0, \infty)$ vanishing at ∞. Its adjoint space is the space $M([0, \infty))$ of all regular complex Borel measures on $[0, \infty)$ with the total variation norm, and

$$\mathcal{L}^* : M([0, \infty)) \to L^\infty([0, \infty))$$

is the Laplace-Stieltjes transform (by a simple application of Fubini's theorem). If the given continuous function h satisfies our lemma's condition, then since

$$||\mathcal{L}\phi||_\infty \leq ||\phi||_1 := ||\phi||_{L^1([0,\infty))},$$

we have necessarily $||h||_\infty \leq M$, i.e., $h \in (L^1)^*$, and by Lemma 2.24, there exists $\mu \in M([0, \infty))$ with $||\mu|| \leq M$ such that $h = \mathcal{L}^*\mu$ (everywhere, by continuity of both sides).||||

2.26. LEMMA. The Laplace-Stieltjes space W for F is a Banach subspace of X, and if $T \in B(X)$ leaves $\mathcal{D}$ invariant and commutes with each $F(s)|_\mathcal{D}$, then $T \in B(W)$ (with $B(W)$ -norm $\leq ||T||$).

PROOF. Clearly, W is a linear manifold in X, and $||.||_W$ is a semi-norm on W. If $x \in W$, then

$$|| \int_0^\infty \phi(t)F(t)xdt|| \leq ||x||_W||\mathcal{L}\phi||_\infty(\leq ||x||_W||\phi||_1) \qquad (1)$$

106

for all $\phi \in C_c^\infty$, hence necessarily

$$\sup_{t \geq 0} ||F(t)x|| \leq ||x||_W \qquad (x \in W). \qquad (2)$$

In particular, since $x = F(0)x$, $||x|| \leq ||x||_W$, and therefore $W := (W, ||.||_W)$ is a normed subspace of X. We prove its completeness. Let $\{x_n\}$ be Cauchy in W (hence in X), and let x be its X-limit. For $\epsilon > 0$ given, let $n_o \in \mathbb{N}$ be such that $||x_n - x_m||_W < \epsilon$ for all $n, m > n_o$. Then for all $\phi \in C_c^\infty$ and $n, m > n_o$,

$$|| \int_0^\infty \phi(t)F(t)(x_n - x_m)dt|| \leq \epsilon||\mathcal{L}\phi||_\infty \leq \epsilon||\phi||_1. \qquad (3)$$

Hence

$$||F(t)(x_n - x_m)|| \leq \epsilon \qquad (n, m > n_o; t \geq 0),$$

that is, $\{F(t)x_n\}$ is uniformly Cauchy in X on $[0, \infty)$. Let then $g(t) := \lim_n F(t)x_n$ (limit in X, uniformly in $t \in [0, \infty)$). Since $x_n \in D(F(t))$ and $x_n \to x$ in X, it follows that $x \in D(F(t))$ and $F(t)x = g(t)$ for each $t \geq 0$, because $F(t)$ is a closed operator. Thus $F(.)x_n \to F(.)x$ uniformly on $[0, \infty)$, so that $F(.)x$ is continuous on $[0, \infty)$ and

$$\int_0^\infty \phi(t)F(t)x_n dt \to \int_0^\infty \phi(t)F(t)x dt \qquad (\phi \in C_c^\infty)$$

strongly in X. Letting $n \to \infty$ in (3), we obtain

$$|| \int_0^\infty \phi(t)F(t)(x - x_m)dt|| \leq \epsilon||\mathcal{L}\phi||_\infty \qquad (m > n_0; \phi \in C_c^\infty).$$

Hence $||x - x_m||_W \leq \epsilon$ for all $m > n_o$; therefore $x - x_m \in W$ (and so $x = (x - x_m) + x_m \in W$), and $||x - x_m||_W \to 0$ when $m \to \infty$. Thus W is complete.

If T is as in the statement of the lemma, then for each $x \in W$, we have $Tx \in D$, $F(.)Tx = T[F(.)x]$ is continuous, and for all $\phi \in C_c^\infty$,

$$|| \int_0^\infty \phi(t)F(t)Tx dt|| = ||T \int_0^\infty \phi(t)F(t)x dt||$$

$$\leq ||T||_{B(X)}||x||_W||\mathcal{L}\phi||_\infty,$$

so that $TW \subset W$ and $||T||_{B(W)} \leq ||T||_{B(X)}.||||$

PROOF OF THEOREM 2.23. For each $x \in W$ and $x^* \in X^*$, $x^*F(.)x$ is a complex continuous function on $[0, \infty)$ satisfying

$$| \int_0^\infty \phi(t)[x^*F(t)x]dt| \leq ||x||_W|||\mathcal{L}\phi||_\infty||x^*|| \qquad (\phi \in C_c^\infty).$$

By Lemma 2.25, there exists a unique $\mu = \mu(.; x, x^*) \in M([0, \infty))$ such that

$$||\mu(.; x, x^*)|| \leq ||x||_W ||x^*|| \qquad (4)$$

and

$$x^* F(t)x = \int_0^\infty e^{-ts} \mu(ds; x, x^*) \qquad (5)$$

for all $t \geq 0$, $x \in W$, and $x^* \in X^*$. The uniqueness of the representation (5) implies the bilinearity of $\mu(\delta; ., .)$ for each fixed $\delta \in \mathcal{B}([0, \infty))$, and since X is reflexive, it follows from (4) that there exists a unique $E(\delta) \in B(W, X)_1$ such that

$$\mu(\delta; x, x^*) = x^* E(\delta)x \qquad (6)$$

for all $x \in W$, $x^* \in X^*$, and $\delta \in \mathcal{B}([0, \infty))$.

Statements (i),(ii),(iii) of the theorem follow now from (6), Pettis' theorem, (5), and Lemma 2.26.

Let (Y, E') be as in the statement of the theorem. Property (ii) for Y contains implicitly the fact that Y is contained in the common domain $\mathcal{D}$ of $F(.)$. Also if $x \in Y$, $F(.)x$ is X-continuous on $[0, \infty)$ (as the Laplace-Stieltjes transform of the vector measure $E'(.)x$), and by Lemma 2.25,

$$||x||_W = \sup\{|\int_0^\infty \phi(t)x^* F(t)x\,dt|; ||x^*|| = 1, \phi \in C_c^\infty, ||\mathcal{L}\phi||_\infty = 1\}$$

$$\leq \sup\{||x^* E'(.)x||; ||x^*|| = 1\} := K_x < \infty,$$

that is, $x \in W$. Since $K_x \leq K||x||_Y$ for a suitable finite constant K, the inclusion $Y \subset W$ is topological. The fact $E'(.) = E(.)|_Y$ follows from the uniqueness property of the Laplace-Stieltjes transform of regular measures. ||||

We shall apply Theorem 2.23 to semigroups of closed operators

2.27. DEFINITION. The family $\{T(t); t \geq 0\}$ of closed operators is called a **semigroup of closed operators** if $T(0) = I$, and $T(s)T(t)x = T(s+t)x$ for all x in the common domain $\mathcal{D}$ of $T(.)$.

2.28. THEOREM. Let $T(.)$ be a semigroup of closed operators on the reflexive Banach space X, and let W be its Laplace-Stieltjes space. Then there exists a uniquely determined spectral measure on W,

$$E : \mathcal{B}([0, \infty)) \to B(W)_1,$$

such that

$$T(t)x = \int_0^\infty e^{-ts} E(ds)x \qquad (t \geq 0; x \in W).$$

Moreover, $E(.)$ commutes with every operator $U \in B(X)$ such that $U\mathcal{D} \subset \mathcal{D}$ and $UT(.)x = T(.)Ux$ for all $x \in \mathcal{D}$. Also, if

$$\tau : h \to \int_0^\infty h(s)E(ds)x,$$

then τ is a norm-decreasing algebra homomorphism of $\mathbb{B}([0,\infty))$ into $B(W)$.

PROOF. Let $(W.E)$ be associated with the family $F(.) = T(.)$ as in Theorem 2.23. Let $x \in W$. Then for each $s \geq 0$, $T(.)T(s)x = T(.+s)x$ is strongly continuous on $[0,\infty)$, and for all $\phi \in C_c^\infty := C_c^\infty(\mathbb{R}^+)$, denoting $\phi_s(u) := \phi(u-s)$, we have

$$\left\| \int_0^\infty \phi(t)T(t)[T(s)x]dt \right\| = \left\| \int \phi_s(u)T(u)x\,du \right\|$$

$$\leq ||x||_W ||\mathcal{L}\phi_s||_\infty \leq ||x||_W ||\mathcal{L}\phi||_\infty,$$

since $(\mathcal{L}\phi_s)(t) = e^{-ts}(\mathcal{L}\phi)(t)$, so that $||\mathcal{L}\phi_s||_\infty \leq ||\mathcal{L}\phi||_\infty$. Thus $T(s)x \in W$, and $||T(s)x||_W \leq ||x||_W$. This shows that $T(.)$ is a semigroup of contractions on the Banach subspace W of X (continuous with respect to the X-norm!).

By Theorem 2.23, for $x \in W$ and $t \geq 0$,

$$T(t) \int_0^\infty e^{-su} E(du)x = T(t)T(s)x$$

$$= T(t+s)x = \int_0^\infty e^{-tu}e^{-su} E(du)x.$$

By linearity, this shows that

$$T(t)\tau(h)x = \int_0^\infty e^{-tu} h(u)E(du)x, \tag{1}$$

for all $x \in W$ and $h(u) = \Sigma_{j=1}^n c_j e^{-s_j u}$ with $c_j \in \mathbb{C}$ and $s_j \geq 0$. These finite linear combinations are dense in $C_b := C_b([0,\infty))$, the space of all bounded continuous complex functions on $[0,\infty)$. If $h \in C_b$, pick a sequence h_k of such combinations such that $h_k \to h$ uniformly on $[0,\infty)$. Then for each $t \geq 0$, $\tau(h_k)x \in D(T(t))$, $\tau(h_k)x \to_k \tau(h)x$, and by (1), $T(t)\tau(h_k)x = \int_0^\infty e^{-tu} h_k(u)E(du)x \to_k \int_0^\infty e^{-tu} h(u)E(du)x$. Since $T(t)$ is closed, it follows that $\tau(h)x \in D(T(t))$ and (1) is valid for all $h \in C_b$. This is easily extended to $h \in \mathbb{B}([0,\infty))$. Indeed, since the vector measure $E(.)x$ is regular (for each $x \in W$), there exists a sequence $\{h_k\} \subset C_b$ such that $||h_k||_\infty = ||h||_\infty$ and $h_k \to h$ pointwise almost everywhere with respect to $E(.)x$. By the Lebesgue Dominated Convergence Theorem for vector measures (cf.[DS-I, p. 328]), $\tau(h_k)x \to \tau(h)x$ in X, $\tau(h_k)x \in D(T(t))$ (for each $t > 0$), and by (1) for h_k, $T(t)\tau(h_k)x \to \int_0^\infty e^{-tu} h(u) E(du) x$. Since $T(t)$ is closed, it follows that $\tau(h)x \in D(T(t))$ and (1) is valid for h.

Thus $\tau(h)x \in \mathcal{D}$, and by (1), we have for all $\phi \in C_c^\infty$,

$$\left\| \int_0^\infty \phi(t) T(t)[\tau(h)x] dt \right\| = \left\| \int_0^\infty \phi(t) \int_0^\infty e^{-tu} h(u) E(du) x \, dt \right\|$$

$$= \left\| \int_0^\infty (\mathcal{L}\phi)(u) h(u) E(du) x \right\| \leq ||h||_\infty ||x||_W ||\mathcal{L}\phi||_\infty.$$

(cf. Theorem 2.23).

Therefore

$$||\tau(h)x||_W \leq ||h||_\infty ||x||_W \qquad (x \in W, h \in \mathbb{B}([0,\infty)), \tag{2}$$

i.e., τ is a norm-decreasing (linear) map of $\mathbb{B}([0,\infty)$ into $B(W)$.

Taking $h = \chi_\delta$, we have $E(\delta) \in B(W)_1$, and by (1), for $x \in W$, etc...,

$$\int_0^\infty e^{-tu} E(du)[E(\delta)x] = T(t)[E(\delta)x] = \int_0^\infty e^{-tu} \chi_\delta(u) E(du) x.$$

By the uniqueness property of the Laplace-Stieltjes transform of regular measures, it follows that
$$E(du) E(\delta) x = \chi_\delta(u) E(du) x,$$

and therefore
$$E(\sigma) E(\delta) x = E(\sigma \cap \delta) x$$

for all $\sigma, \delta \in \mathcal{B}([0,\infty))$ and $x \in W$. Thus E is a spectral measure on W, and τ is necessarily multiplicative on $\mathbb{B}([0,\infty))$, since it is multiplicative on the simple Borel functions, and satisfies (2) on $\mathbb{B}([0,\infty))$.

In view of Theorem 2.23, this completes the proof of Theorem 2.28.||||

In the special case of a C_o-semigroup of contractions $T(.)$, with generator $-A$, since $\mathbb{R}^+ \subset \rho(-A)$, the semi-simplicity manifold Z for $-A$ is well-defined, as well as the Laplace-Stieltjes space W for $T(.)$. As expected, we have

2.29. THEOREM. Let $-A$ generate the C_o-semigroup of contractions $T(.)$ on the reflexive Banach space X. Let Z and W be the semi-simplicity manifold for A and the Laplace-Stieltjes space for $T(.)$, respectively. Then $Z = W$, topologically.

PROOF. The observations preceding Definition 2.22 and the maximality of W show that $Z \subset W$.

On the other hand, if $x \in W$, then

$$R(t)x = \int_0^\infty e^{-ts}T(s)x \, ds = \int_0^\infty e^{-ts} \int_0^\infty e^{-su}E(du)x \, ds$$

$$= \int_0^\infty \int_0^\infty e^{-(t+u)s} ds \, E(du)x = \int_0^\infty \frac{1}{t+u} E(du)x,$$

where the change of integration order is easily justified by the Tonnelli and Fubini theorems. By the multiplicativity of the map τ induced by $E(.)$, the spectral measure on W, we get for all $k \in \mathbb{N}$,

$$S^k(t)x := \{tR(t)[1 - tR(t)]\}^k x = \int_0^\infty \{\frac{t}{t+u}[1 - \frac{t}{t+u}]\}^k E(du)x$$

$$= \int_0^\infty \{\frac{tu}{(t+u)^2}\}^k E(du)x = \int_0^\infty (\frac{u}{t})^k (1 + \frac{u}{t})^{-2k} E(du)x.$$

Therefore, for all unit vectors $x^* \in X^*$,

$$\|x^* S^k x\|_1 \le \int_0^\infty \int_0^\infty (\frac{u}{t})^k (1 + \frac{u}{t})^{-2k} \frac{dt}{t} |x^* E(.)x|(du)$$

$$= \int_0^\infty \int_0^\infty s^k (1 + s)^{-2k} \frac{ds}{s} |x^* E(.)x|(du)$$

$$= B(k, k)\|x^* E(.)x\| \le B(k, k)\|x\|_W.$$

Hence

$$\|x\|_Z := \sup\{\|x\|, \frac{\|x^* S^k x\|_1}{B(k, k)}; k \in \mathbb{N}, \quad \|x^*\| = 1\} \le \|x\|_W < \infty,$$

that is, $W \subset Z$.

We thus proved that $Z = W$, and since both are Banach spaces and $\|.\|_Z \le \|.\|_W$, it follows that the norms are equivalent (by a well-known theorem of Banach). ‖‖‖

We consider now a "variation" of the construction of the Laplace-Stieltjes space for a family $F(.)$ as given in Definition 2.22. In that construction, the constraint on $\phi \in C_c^\infty := C_c^\infty(\mathbb{R}^+)$ was $\|\mathcal{L}\phi\|_\infty = 1$. Since ϕ vanishes in a neighborhood of 0, $\mathcal{L}\phi \in L^1(\mathbb{R}^+, dt)$ (with the *usual* Lebesgue measure dt), and has $L^1(\mathbb{R}^+, dt)$-norm

$$\|\mathcal{L}\phi\|_1 \le \|\phi\|_{L^1(\mathbb{R}^+, ds/s)}.$$

It then makes sense to replace the norm $\|.\|_\infty$ by the norm $\|.\|_1$ in the constraint on ϕ (in Definition 2.22).

2.30. DEFINITION. Let $\{F(t); t \geq 0\}$ be a family of closed operators with common domain $\mathcal{D}$. The "Integrated Laplace space" for $F(.)$ is the set Y of all $x \in \mathcal{D}$ such that $F(.)x$ is X-continuous on $\mathbb{R}^+$, and

$$||x||_Y := \sup\{||x||, \left|\left| \int_0^\infty \phi(t)F(t)x\,dt \right|\right|; \phi \in C_c^\infty, ||\mathcal{L}\phi||_1 = 1\}$$

is finite.

2.31. THEOREM. Let Y be the Integrated Laplace space for the family $F(.)$ of closed operators on the *arbitrary* Banach space X. Then Y is a Banach subspace of X, invariant for every $T \in B(X)$ commuting with $F(t)|_\mathcal{D}$ for all $t > 0$ (and $||T||_{B(Y)} \leq ||T||_{B(X)}$), and there exists a uniquely determined map

$$S(.) : [0, \infty) \to B(Y, X)$$

with the following properties: (1) $S(0) = 0$;
(2) $||S(t)x - S(u)x|| \leq |t - u|\,||x||_Y$ $(t, u \geq 0; x \in Y)$;
(3) $F(t)x = t \int_0^\infty e^{-tu} S(u)x\,du$ $(t > 0; x \in Y)$.
Moreover, the pair (Y,S) is "maximal-unique" in the usual sense.

PROOF. The basic properties of Y are verified in precisely the same way as the corresponding properties of the Laplace-Stieltjes space W.

Denote by Lip_o the space of all complex functions f on $[0, \infty)$ such that $f(0) = 0$ and

$$|f(t) - f(u)| \leq M|t - u| \qquad (t, u \geq 0).$$

The smallest possible constant M above is called the Lipshitz constant for the "Lipshitz function" f.

The remainder of the proof depends on the following

LEMMA. Let $h : \mathbb{R}^+ \to \mathbb{C}$ and $K > 0$ be given. Then there exists $f \in Lip_o$ with Lipshitz constant $\leq K$ such that $h(t)/t$ is the Laplace transform of f on $\mathbb{R}^+$ if and only if h is continuous and

$$\left| \int_0^\infty \phi(t)h(t)dt \right| \leq K||\mathcal{L}\phi||_1 \tag{*}$$

for all $\phi \in C_c^\infty$.

PROOF (of Lemma). If $h(t)/t = (\mathcal{L}f)(t)$ for all $t > 0$, where $f \in Lip_o$ has Lipshitz constant $\leq K$, then h is clearly continuous, and f is locally absolutely continuous, its derivative (which exists a.e.) satisfies $||f'||_\infty \leq K$, and

$$f(t) = (Jf')(t) := \int_0^t f'(s)ds \qquad (t \geq 0).$$

For any $\phi \in C_c^\infty$, integration by parts and Fubini's theorem give

$$\int_0^\infty \phi(t)h(t)dt = \int_0^\infty \int_0^\infty te^{-tu}(Jf')(u)du\phi(t)dt$$

$$= \int_0^\infty \int_0^\infty e^{-tu}f'(u)du\phi(t)dt = \int_0^\infty (\mathcal{L}\phi)(u)f'(u)du.$$

Thus

$$\left| \int_0^\infty \phi(t)h(t)dt \right| \leq ||f'||_\infty ||\mathcal{L}\phi||_1 \leq K||\mathcal{L}\phi||_1$$

for all $\phi \in C_c^\infty$.

Conversely, suppose h is continuous on $\mathbb{R}^+$ and satisfies $(*)$.
For any $\phi \in C_c^\infty := C_c^\infty(\mathbb{R}^+)$,

$$||\mathcal{L}\phi||_1 \leq \int_0^\infty \int_0^\infty te^{-tu}du|\phi(t)|dt/t \leq ||\phi||_{L^1(\mathbb{R}^+,dt/t)}.$$

That is, the operator

$$\mathcal{L} : L^1(dt/t) := L^1(\mathbb{R}^+, dt/t) \to L^1(dt) := L^1(\mathbb{R}^+, dt) \qquad (1)$$

is a contraction.

We identify $[L^1(dt/t)]^*$ with the space of all complex measurable functions h on $\mathbb{R}^+$ such that $th(t) \in L^\infty := L^\infty(\mathbb{R}^+, dt)$, normed by the essential supremum norm of $th(t)$, with the duality given by

$$< \phi, h > = \int_0^\infty \phi(t)h(t)dt = \int_0^\infty \phi(t)[th(t)](dt/t).$$

By Fubini's theorem, for all $\phi \in L^1(dt/t)$ and $\psi \in L^\infty$,

$$< \mathcal{L}\phi, \psi > = \int_0^\infty (\mathcal{L}\phi)(s)\psi(s)ds = \int_0^\infty (\mathcal{L}\psi)(t)\phi(t)dt.$$

The use of Fubini's theorem is justified because

$$\int \int e^{-st}|\phi(t)|.|\psi(s)|dtds \leq ||\psi||_\infty ||\phi||_{L^1(dt/t)} < \infty.$$

This shows that the operator $\mathcal{L}$ defined in (1) has the adjoint

$$\mathcal{L}^* = \mathcal{L} : L^\infty(dt) \to [L^1(dt/t)]^*.$$

By (∗) for h,

$$\left| \int_0^\infty [th(t)][\phi(t)/t]dt \right| \le K||\mathcal{L}\phi||_1 \le K||\phi||_{L^1(dt/t)}$$

for all $\phi \in C_c^\infty$, and therefore $||th(t)||_\infty \le K$. This means that $h \in [L^1(dt/t)]^*$, and (∗) is precisely Condition (∗) in Lemma 2.24 for the operator $T = \mathcal{L}$. There exists therefore $\psi \in L^\infty(dt)$ with $||\psi||_\infty \le K$, such that $h = \mathcal{L}\psi$ (everywhere on $\mathbb{R}^+$, by continuity of both sides). Now $f := J\psi \in Lip_o$, with Lipshitz constant $||\psi||_\infty \le K$, and an integration by parts shows that $h(t) = t\int_0^\infty e^{-ts}f(s)ds$.||||

PROOF OF THEOREM 2.31. Fix $x \in Y$ and $x^* \in X^*$. The function $h := x^*F(.)x$ satisfies the criterion in the lemma with $K = ||x||_Y||x^*||$. There exists therefore a unique function $f = f(.; x, x^*) \in Lip_o$, such that
(1) $x^*F(t)x = t\int_0^\infty e^{-tu}f(u; x, x^*)du$ $(t > 0)$; and
(2) $|f(t; x, x^*) - f(u; x, x^*)| \le |t - u|.||x||_Y||x^*||$ $(t, u \ge 0)$.
In particular (for $u = 0$)

$$|f(t; x, x^*)| \le |t|.||x||_Y||x^*|| (t \ge 0; x \in Y, x^* \in X^*).$$

The uniqueness of the representation (1) implies that $f(t; ., .)$ is a bounded bilinear form (for each fixed t), and there exists therefore a uniquely determined operator $S(t) \in B(Y, X^{**})$ such that
(3) $f(t; x, x^*) = [S(t)x](x^*)$ for all $x \in Y$ and $x^* \in X^*$,
and by (2),
(4) $||S(t)x - S(u)x||_{X^{**}} \le |t - u|.||x||_Y$ for all $t, u \ge 0$ and $x \in Y$.
For $t = 0$, the left side of (3) vanishes for all x, x^*, and therefore $S(0) = 0$.
By (4), the integral $\int_0^\infty e^{-tu}S(u)xdu$ (with $t > 0$ and $x \in Y$) converges strongly in X^{**}, and we may then rewrite (1) in the form

$$\kappa[F(t)x] = t\int_0^\infty e^{-tu}S(u)xdu,$$

where κ denotes the canonical imbedding of X into X^{**}.
Let π denote the canonical homomorphism

$$\pi : X^{**} \to X^{**}/\kappa X.$$

Since π is continuous and $\pi\kappa = 0$, we obtain

$$0 = \pi\kappa[F(t)x] = t\int_0^\infty e^{-tu}\pi[S(u)x]du   (t > 0).$$

The uniqueness of the Laplace transforms implies that $\pi[S(u)x] = 0$, i.e., $S(u)x \in \kappa X$ for all $u \geq 0$ and $x \in Y$. Identifying as usual κX with X, we may restate the above relations as the statements (1)-(3) of the theorem.

The maximal-uniqueness is an immediate consequence of the necessity part of the lemma and the uniqueness of the Laplace transform.‖‖‖

Note that Statement (2) in Theorem 2.31 means that $S(.)$ is of class $Lip_{0,1}$ as a $B(Y,X)$-valued function (where the index 1 indicates that the Lipshitz constant is ≤ 1), that is,

$$\|S(t) - S(u)\|_{B(Y,X)} \leq |t - u| \qquad (t, u \geq 0).$$

In particular, the Laplace transform $\mathcal{L}S$ exists in the $B(Y,X)$-norm on $(0, \infty)$, and by Statement (3) of the theorem, $F(.)$ is $B(Y,X)$-valued and $F(t) = t(\mathcal{L}S)(t)$ for all $t > 0$.

We express this relation between $F(.)$ and $S(.)$ by saying that $F(.)$ is the integrated Laplace transform of the $B(Y,X)$- valued $Lip_{0,1}$-function $S(.)$.

2.32. COROLLARY. Let $F(.)$ be a family of closed operators on $(0, \infty)$, operating on the arbitrary Banach space X, and let $Y := (Y, \|.\|_Y)$ be its Integrated Laplace space. Then the following statements are equivalent.

(1) $Y = X$.
(2) $K := \sup_{\|x\|=1} \|x\|_Y < \infty$.
(3) $F(.)$ is the integrated Laplace transform of a $B(X)$-valued $Lip_{0,K}$-function.

In view of the characterization of semigroups generators given in Lemma 5 (in the proof of Theorem 1.38), it is interesting to consider the special family $F(t) = R(t; A)$ for a given operator A.

2.33. DEFINITION. The operator A on the Banach space X, with $(0, \infty) \subset \rho(A)$ is said to generate an **integrated semigroup of bounded type** $\leq K$ if $R(.; A)$ is the integrated Laplace transform of a $B(X)$-valued $Lip_{0,K}$-function $S(.)$ on $[0, \infty)$.

The (uniquely determined) function $S(.)$ is called the **integrated semigroup generated by** A.

By Corollary 2.32, we have

2.34. COROLLARY. An operator A with $(0, \infty) \subset \rho(A)$, acting in a Banach space X, is the generator of an integrated semigroup of bounded type $\leq K$ if and only if

$$\left\| \int_0^\infty \phi(t) R(t; A) dt \right\| \leq K \|\mathcal{L}\phi\|_1$$

for all $\phi \in C_c^\infty(\mathbb{R}^+)$.

Recall that $\|.\|_1$ denotes the $L^1(\mathbb{R}^+, dt)$-norm.

The more general case where $(a, \infty) \subset \rho(A)$ for some $a \geq 0$ and $S(.)$ is of exponential type $\leq a$ is easily reduced to the case above by translation. We omit the details.

If $n \in \mathbb{N}$ is given, the operator A (with $(0, \infty) \subset \rho(A)$) generates an **n-times integrated semigroup of bounded type** $\leq K$ if $R(t; A) = t^n (\mathcal{L}S)(t)$ on $(0, \infty)$, for S as in Definition 2.33, i.e., if $t^{-(n-1)} R(t; A)$ is the integrated Laplace transform of S. These objects have been studied recently, and have been found to be useful in the analysis of the Abstract Cauchy Problem. We refer to the bibliography for additional information. Let us only state the following immediate consequence of Corollary 2.32 and of the above observation:

2.35. COROLLARY. The operator A with $(0, \infty) \subset \rho(A)$ generates an n-times integrated semigroup of bounded type $\leq K$ if and only if

$$\left\| \int_0^\infty \phi(t) R(t; A) \frac{dt}{t^{n-1}} \right\| \leq K \|\mathcal{L}\phi\|_1$$

for all $\phi \in C_c^\infty(\mathbb{R}^+)$.

There is a simple relation between exponentially bounded n-times integrated semigroups and (exponentially bounded) pre-semigroups. In order to simplify the statement, we consider only the case $n = 1$ and we assume that $[0, \infty) \subset \rho(A)$ (making a translation as needed).

2.36. THEOREM. Let A be such that $[0, \infty) \subset \rho(A)$. Then A generates the integrated semigroup $S(.)$ (of bounded type $\leq K$) if and only if it generates the pre-semigroup $A^{-1} + S(.)$ (of class $Lip_{A^{-1}, K}$).

PROOF. The characterization in Lemma 5 (in the proof of Theorem 1.38) carries over to pre-semigroups in the following form:

The operator A (with $[0, \infty) \subset \rho(A)$) generates the pre-semigroup $W(.)$ of class $Lip_{A^{-1}}$ if and only if

$$A^{-1} R(t; A)x = \int_0^\infty e^{-tu} W(u)x \, du$$

for all $t > 0, x \in X$.

Writing briefly $\mathcal{L}W$ for the above Laplace transform (understood in the strong operator topology), since $\mathcal{L}A^{-1} = t^{-1} A^{-1}$, the above condition is equivalent to

$$A^{-1}[-t^{-1} + R(t; A)] = \mathcal{L}(-A^{-1} + W)(t) \qquad (t > 0),$$

that is,

$$A^{-1}[-I + tR(t; A)] = t\mathcal{L}(-A^{-1} + W)(t),$$

i.e.,

$$R(t; A) = t(\mathcal{L}S)(t) \qquad (t > 0),$$

where $S := -A^{-1} + W$. ||||

E. SEMIGROUPS OF UNBOUNDED SYMMETRIC OPERATORS

In this section, Stone's theorem is generalized to semigroups of *unbounded* symmetric operators.

Let $\Delta = [0, c]$ $(c > 0)$, and let $\{T(t); t \in \Delta\}$ be a family of *unbounded* operators acting in a *Hilbert* space X, with $T(0) = I$ and $D(T(t)) := D_t$ $(t \in \Delta)$. *We assume that* $D_s \subset D_t$ for $s \geq t$, and that the linear manifold

$$D := \bigcup \{D_t; 0 < t \leq c\}$$

is *dense* in X.

The semigroup condition takes domains into account as follows:
for $t, s, t + s \in \Delta$, we have $T(s)D_{t+s} \subset D_t$, and

$$T(t)T(s) = T(t + s) \quad on \quad D_{t+s}.$$

The (weak) continuity condition takes the following form:
for each $0 < s \in \Delta$, $(T(.)x, x)$ is continuous on $[0, s]$, for all $x \in D_s$.
Briefly, the above family of operators will be called a *local semigroup* (on Δ). It is *symmetric* if each $T(t)$ is a symmetric operator $(t \in \Delta)$, that is

$$(T(t)x, y) = (x, T(t)y) \qquad (x, y \in D_t, t \in \Delta).$$

2.37. THEOREM. Let $T(.)$ be a symmetric local semigroup on Δ. Then there exists a unique selfadjoint operator H such that for all $0 < s \in \Delta$, $D_s \subset D(e^{-tH})$ and $T(t) = e^{-tH}$ on D_s for all $t \in [0, s/2]$.

PROOF. *Fix* $0 < s \in \Delta$ and $x \in D_s$, and consider the *continuous* function

$$f(t) := (T(t)x, x) \qquad (t \in D_s).$$

For any $n \in \mathbb{N}$, if $t_1, ..., t_n \in [0, s]$ are such that $t_i + t_j \in [0, s]$, then $x \in D_{t_i}$, $T(t_i)x \in D_{s-t_i} \subset D_{t_j}$ (since $s - t_i \geq t_j$), and $T(t_j)T(t_i)x = T(t_i + t_j)x$. Therefore, by the symmetry hypothesis,

$$f(t_i + t_j) = (T(t_i)x, T(t_j)x).$$

If now $c_1, ..., c_n \in \mathbb{C}$, then

$$\Sigma_{i,j=1}^{n} c_i c_j^- f(t_i + t_j) = \Sigma_{i,j} c_i c_j^- (T(t_i)x, T(t_j)x) = ||\Sigma_i c_i T(t_i)x||^2 \geq 0.$$

By a theorem of Widder [W1], this positivity property of the continuous function f on $[0, s]$ implies the existence of a unique regular positive Borel measure $\alpha = \alpha(.; x)$ on $\mathbb{R}$ such that $e^{-tu} \in L^1(\mathbb{R}, \alpha)$ and

$$f(t) = \int_{\mathbb{R}} e^{-tu} \alpha(du) \qquad (t \in [0, s]). \tag{1}$$

For $n, m \in \mathbb{N}$, $c_1, ..., c_n; d_1, ..., d_m$ complex, and $s_1, ..., s_n; t_1, ..., t_m$ real such that $s_i + t_j \in [0, s]$, the preceding calculation shows that

$$(\Sigma_{i=1}^{n} c_i T(s_i)x, \Sigma_{j=1}^{m} d_j T(t_j)x) = \Sigma_{i,j} c_i d_j^- (T(s_i)x, T(t_j)x)$$

$$= \Sigma_{i,j} c_i d_j^- f(s_i + t_j) = \Sigma_{i,j} c_i d_j^- \int_{\mathbb{R}} e^{-(s_i + t_j)u} \alpha(du)$$

$$= \int_{\mathbb{R}} (\Sigma_i c_i e^{-s_i u})(\Sigma_j d_j e^{-t_j u})^- \alpha(du). \tag{2}$$

Let Y be the closed span of $\{T(t)x; t \in [0, s/2]\}$, and let $U(T(t)x) := e^{-tu} (\in L^2(\alpha)$, since $e^{-2tu} \in L^1(\alpha)$ for $t \in [0, s/2]$). If $g \in L^2(\alpha)$ is orthogonal to all the functions e^{-tu} with $t \in [0, s/2]$, then the function $G(z) := \int_{\mathbb{R}} e^{-zu} g^-(u)\alpha(du)$, which is analytic in the strip $S := \{z \in \mathbb{C}; Rez \in (0, s/2)\}$ and continuous in its closure S^-, must vanish identically. Hence $\int_{\mathbb{R}} e^{-iru} g(u)^- \alpha(du) = 0$ for all real r. By the uniqueness property of the Fourier-Stieltjes transform, it follows that $g(u)^- \alpha(du) = 0$, and therefore $\int_{\mathbb{R}} g g^- d\alpha = 0$, i.e., g is the zero element of $L^2(\alpha)$. Thus $\{e^{-tu}; t \in [0, s/2]\}$ is fundamental in $L^2(\alpha)$, and it follows from (2) that U extends linearly to a unitary operator from Y onto $L^2(\alpha)$.

For each $z \in S^-$, the function $h_z(u) = e^{-zu}$ is in $L^2(\alpha)$. The $L^2(\alpha)$-valued function $z \rightarrow h_z$ is continuous in S^-. Indeed, let $\phi(u) = 1$ for $u \geq 0$ and $\phi(u) = e^{-su}$ for $u < 0$. Let $z, w \in S^-$, and denote $Rez = t$, $Rew = r$. Then

$$|e^{-zu} - e^{-wu}|^2 \leq (e^{-tu} + e^{-ru})^2$$

$$= e^{-2tu} + e^{-2ru} + 2e^{-(t+r)u} \leq 4\phi(u) \in L^1(\alpha),$$

since $2t, 2r, t + r \leq s$. It then follows by dominated convergence that

$$||h_z - h_w||^2_{L^2(\alpha)} \rightarrow 0$$

when $w \rightarrow z$.

For each $g \in L^2(\alpha)$, $(h_z, g) = \int_{\mathbb{R}} e^{-zu} g(u)^- \alpha(du)$ is the Laplace-Stieltjes transform of the measure $g^- d\alpha$; it converges absolutely in S^- since $e^{-tu} \in L^2(\alpha)$ for $t \in [0, s/2]$, and is therefore analytic in S. Hence the $L^2(\alpha)$-valued function h_z is analytic in S.

Define
$$x(z) := U^{-1} h_z (\in Y) \qquad (z \in S^-). \qquad (3)$$

Then, as a Y-valued function, $x(.)$ is continuous in S^-, analytic in S, and
$$x(t) = T(t)x \qquad (t \in [0, s/2]). \qquad (4)$$

Now, given $x \in D$, there exists $0 < s \in \Delta$ such that $x \in D_s$. Let $x(.)$ be the function constructed above in the strip S^-, and define
$$V(r)x := x(ir) \qquad (r \in \mathbb{R}). \qquad (5)$$

By (4) and the analyticity of $x(.)$, $V(.)$ is well-defined on D. For $r, r' \in \mathbb{R}$, we have
$$(V(r)x, V(r')x) = (x(ir), x(ir')) = (h_{ir}, h_{ir'})$$
$$= \int_{\mathbb{R}} e^{-i(r-r')u} \alpha(du). \qquad (6)$$

Thus, by (1),
$$\|V(r)x\|^2 = \alpha(\mathbb{R}) = f(0) = \|x\|^2 \qquad (7)$$

for all $r \in \mathbb{R}$ and $x \in D$, i.e., each $V(r)$ is an isometry from D to X. We verify its linearity as follows. If $x, y \in D$ and $\lambda, \mu \in \mathbb{C}$, there exists $0 < s \in \Delta$ such that $x, y \in D_s$. Then by (4), for all $t \in [0, s/2]$,
$$(\lambda x + \mu y)(t) = T(t)(\lambda x + \mu y) = \lambda T(t)x + \mu T(t)y = \lambda x(t) + \mu y(t),$$

and therefore, by analyticity of $x(.)$ and $y(.)$ in S and their continuity in S^-, the same relation is valid with t replaced by ir, i.e., $V(r)(\lambda x + \mu y) = \lambda V(r)x + \mu V(r)y$ for all real r.

Since $V(r)$ is isometric on the *dense* linear manifold D, it extends as a linear isometry on X.

The function $r \to V(r)x = x(ir)$ is continuous for each $x \in D$ (as observed above), and $V(r)$ is isometric on X; therefore $V(.)x$ is continuous on $\mathbb{R}$ for *all* $x \in X$.

Let $x \in D_s$ and $t, t' \in \Delta$ such that $t+t' \in \Delta$. The semigroup property $T(t+t')x = T(t)T(t')x$ implies, by uniqueness of the analytic continuation onto the imaginary axis, that $V(r)V(r')x = V(r+r')x$ for all $r, r' \in \mathbb{R}$ and $x \in D_s$, hence for all $x \in X$, by density of D. Thus $V(.)$ is a group of operators on $\mathbb{R}$; in particular, the isometries $V(r)$ are *onto*, i.e., $V(.)$ is a strongly continuous *unitary* group.

By Stone's theorem, we have $V(r) = e^{-irH}$ with H *selfadjoint*. Let E be the resolution of the identity for H. For all $r \in \mathbb{R}$ and $x \in D_s$, we have by (6)

$$\int_{\mathbb{R}} e^{-iru}(E(du)x, x) = (V(r)x, x) = \int_{\mathbb{R}} e^{-iru} \alpha(du; x),$$

and therefore $(E(.)x, x) = \alpha(.; x)$, by the uniqueness property of the Fourier-Stieltjes transform.

Since $e^{-zu} \in L^2(\alpha)$ for $z \in S^-$, we have $x \in D(e^{-zH})$. The vector functions $e^{-zH}x$ and $x(z)$ are both analytic in S and continuous in S^-; on the imaginary axis, we have $e^{-irH}x = V(r)x = x(ir)$, so that $e^{-zH}x = x(z)$ for all $z \in S^-$. In particular, by (4),

$$T(t)x = e^{-tH}x \qquad (t \in [0, s/2], x \in D_s). \tag{8}$$

If also H' is a selfadjoint operator satisfying (8), then the analytic continuation employed in the construction gives $e^{-irH}x = e^{-irH'}x$ for all real r and all $x \in D$, hence for all $x \in X$, and therefore $H = H'$ by the uniqueness in Stone's theorem.||||

We apply Theorem 2.37 to "analytic vectors".

2.38. DEFINITION. Let A be an (unbounded) operator on X. An **analytic vector** for A is a vector $x \in D^\infty(A)$ such that, for some $t > 0$ (depending on x),

$$\Sigma_{n=0}^\infty \frac{t^n}{n!}||A^n x|| < \infty.$$

2.39. THEOREM (Nelson's Analytic Vectors Theorem). Let A be a closed symmetric operator on a Hilbert space X, that possesses a dense set D of analytic vectors. Then A is *selfadjoint*.

PROOF. For $t > 0$, let

$$D_t = \{x \in D^\infty(A); \Sigma \frac{t^n}{n!}||A^n x|| < \infty\}.$$

These are linear manifolds such that $D_s \subset D_t$ for $s \geq t$ and $\bigcup_{t>0} D_t = D$ is dense, by hypothesis. Define $T(0) = I$ and

$$T(t)x = \Sigma_{n=0}^\infty \frac{t^n}{n!} A^n x$$

for $x \in D_t, t > 0$ (the series converges in X for such x).

120

$\in D_{t+s}$,

$$\Sigma_m \Sigma_n \frac{t^m s^n}{m!n!} \|A^{m+n}x\| = \Sigma_k (1/k!) \Sigma_{m=0}^{k} \binom{k}{m} t^m s^{k-m} \|A^k x\|$$

$$= \Sigma_k \frac{(t+s)^k}{k!} \|A^k x\| < \infty. \tag{9}$$

Thus $\Sigma_n \frac{s^n}{n!}\|A^n(A^m x)\| < \infty$, that is, $A^m x \in D_s$ for all $m = 0, 1, 2, ...$, and since A is closed, one verifies by induction that $T(s)x \in D(A^m)$, and

$$T(s)A^m x = A^m T(s)x = \Sigma_n \frac{s^n}{n!} A^{n+m} x. \tag{10}$$

Therefore

$$\Sigma_m \frac{t^m}{m!} \|A^m T(s)x\| = \Sigma_m \frac{t^m}{m!} \|\Sigma_n \frac{s^n}{n!} A^{m+n} x\| < \infty$$

by (9), i.e., $T(s)x \in D_t$, and by (10) and absolute convergence,

$$T(t)T(s)x = \Sigma_m \frac{t^m}{m!} \Sigma_n \frac{s^n}{n!} A^{n+m} x$$

$$= \Sigma_k \frac{(t+s)^k}{k!} A^k x = T(t+s)x$$

(for all $t, s \geq 0$ and $x \in D_{t+s}$).

For $s > 0$ and $x \in D_s$, we have for all $t \in [0, s]$

$$(T(t)x, x) = \Sigma_n \frac{t^n}{n!}(A^n x, x),$$

where the series converges absolutely; in particular, $(T(.)x, x)$ is continuous on $[0, s]$.

By symmetry of A, we clearly have $(T(t)x, y) = (x, T(t)y)$ for all $x, y \in D_t$.

We conclude that $\{T(t); t \geq 0\}$ is a *symmetric local semigroup*. Let H be the *selfadjoint* operator associated with it as in Theorem 2.37. If $x \in D$, then $x \in D_s$ for some $s > 0$, and

$$e^{-tH} x = T(t)x := \Sigma_n \frac{t^n}{n!} A^n x \tag{11}$$

for all $t \in [0, s/2]$.

In particular $e^{-tH} x \in D_{s-t} \subset D$ (for $t \in [0, s/2]$), and it follows that D is *invariant* for the C_o-(semi)group e^{-tH}. Since D is also dense in X by hypothesis, Theorem 1.7 implies that D is a *core* for the generator H, if $D \subset D(H)$.

However for $x \in D$, we have by (11)

$$Hx := \lim_{t \to 0+} t^{-1}[e^{-tH} x - x] = Ax,$$

i.e., $x \in D(H)$ and $Hx = Ax$.

Thus indeed $D \subset D(H)$, $Hx = Ax$ for all $x \in D$, and D is a core for H. Since A is closed, we have

$$H = (H/D)^- = (A/D)^- \subset A^- = A,$$

and since A is symmetric and H is selfadjoint, it follows that

$$A \subset A^* \subset H^* = H,$$

i.e., $A = H$, so that A is indeed selfadjoint.||||

If A is not assumed to be closed, the theorem applies to its closure A^-, which is closed and symmetric, and every analytic vector for A is certainly an analytic vector for A^-. We then have

2.40. COROLLARY. Let A be a symmetric operator with a dense set of analytic vectors. Then A is essentially selfadjoint.

Semigroups of operators are associated with the ACP

$$u' = Au \qquad u(0) = x.$$

The **second order ACP**

$$u'' = Au \qquad u(0) = x, \quad u'(0) = 0$$

in Banach space is associated in a similar way to so-called "cosine operator functions"(cf. [G]).

2.41. DEFINITION. A **cosine operator function** on the Banach space X is a function $C(.) : \mathbb{R} \to B(X)$ such that $C(0) = I$ and

$$C(t + s) + C(t - s) = 2C(t)C(s) \qquad (t, s \in \mathbb{R}).$$

The concept that parallels that of a *local* semigroup is the following

2.42. DEFINITION. Let D be a *dense* linear manifold in X. A **local cosine family** (of operators) on D is a family $\{C(t); t \in \mathbb{R}\}$ of operators on the Banach space X, such that for each $x \in D$ there exists $\epsilon = \epsilon(x) > 0$ such that
(i) $x \in D(C(t))$ and $C(.)x$ is strongly continuous for $|t| < \epsilon$;
(ii) $C(0)x = x$, and for $|t|, |s|, |t + s|, |t - s| < \epsilon$, $C(s)x \in D(C(t))$ and

$$C(t + s)x + C(t - s)x = 2C(t)C(s)x.$$

A result parallel to Theorem 2.37 for local cosine families of symmetric operators in Hilbert space is stated below, first for the special case when all the operators $C(t)$ are *bounded below*, that is,

$$(C(t)x, x) \geq ||x||^2 \qquad (x \in D(C(t)), t \in \mathbb{R}). \tag{1}$$

Condition (1) implies in particular that all the operators $C(t)$ are *symmetric*. The general case of a *symmetric* local cosine family is dealt with in Theorem 2.45.
Since no parallel to Widder's theorem [W1] is known for the cosine transform, the proof will proceed differently.

2.43. THEOREM. Let D be a dense linear manifold in the (complex) Hilbert space X, and let $C(.)$ be a local cosine family of bounded below operators on D. Then there exists a unique positive selfadjoint operator A such that

$$C(t)x = \cosh(tA^{1/2})x$$

for all $x \in D$ and $|t| < \epsilon(x)$.

Note that the family $\{\cosh(tA^{1/2}); t \in \mathbb{R}\}$ is a cosine family of bounded below *selfadjoint* operators that *extends* the local family $C(.)$.

PROOF. Since $C(t)$ is *symmetric* for each t, it is closable, and its closure $C(t)^-$ clearly satisfies (i), (ii), and (1). We may then assume that $C(.)$ is a local cosine family of *closed* bounded below operators on D, replacing $C(t)$ by $C(t)^-$ if needed (by (i), the conclusion of the theorem remains unchanged).

Fix a sequence $\{h_n\}$ of non-negative C^∞- functions on $\mathbb{R}$, such that $h_n(t) = 0$ for $|t| \geq 1/n$ and $\int_{\mathbb{R}} h_n(t)dt = 1$.

Let $x \in D$, and *fix* $n(x) > 1/\epsilon(x)$. Denote

$$x_n = \int_{\mathbb{R}} h_n(s)C(s)x\,ds \qquad (n \geq n(x)), \tag{2}$$

where the integral is a well-defined strong integral, by (i) in Definition 2.42.

Clearly $x_n \to x$ strongly. Since D is dense by hypothesis, it follows that the set

$$D_0 := \{x_n; x \in D, n \geq n(x)\}$$

is *dense* in X.

Fix $n \geq n(x)$. If $|t| < \epsilon_n(x) := \epsilon(x) - 1/n$ (note that $1/n \leq 1/n(x) < \epsilon(x)$), Condition (ii) implies that $C(s)x \in D(C(t))$ for all s with $|s| < 1/n$. Also $C(t)C(s)x = (1/2)[C(t+s)x + C(t-s)x]$ is strongly continuous for $|s| < 1/n$ (because $|t+s|, |t-s| < \epsilon(x)$). Since $C(t)$ is closed, it follows from Theorem 3.3.2 in [HP] that

$$x_n \in D(C(t))$$

and

$$C(t)x_n = \int_{\mathbb{R}} h_n(s)C(t)C(s)x\,ds \tag{3}$$

$(|t| < \epsilon_n(x))$. Let $u > 0$ be such that $|t+u|, |t-u| < \epsilon_n(x)$ (for a given t such that $|t| < \epsilon_n(x)$). By (3)

$$[C(t+u) + C(t-u) - 2C(t)]x_n$$

$$= \int_{\mathbb{R}} h_n(s)[C(t+u) + C(t-u) - 2C(t)]C(s)x\,ds$$

$$= \int_{\mathbb{R}} h_n(s)C(s)[...]x ds = \int_{\mathbb{R}} h_n(s)C(s)[2C(t)C(u) - 2C(t)]x ds$$

$$= \int_{\mathbb{R}} h_n(s)C(t)[2C(s)C(u) - 2C(s)]x ds$$

$$= \int_{\mathbb{R}} h_n(s)C(t)[C(s+u) + C(s-u) - 2C(s)]x ds$$

$$= \int_{\mathbb{R}} [h_n(v-u) + h_n(v+u) - 2h_n(v)]C(t)C(v)x dv.$$

In the last integral, integration extends over an interval where $|t|, |v|, |t+v|, |t-v| < \epsilon(x)$, so that Conditions (i),(ii) imply that $C(t)C(v)x$ is strongly continuous there (as a function of v), and therefore

$$u^{-2}[C(t+u) + C(t-u) - 2C(t)]x_n \to_{u \to 0} \int_{\mathbb{R}} h_n''(v)C(t)C(v)x dv \qquad (4)$$

strongly (for $|t| < \epsilon_n(x)$).

Let

$$D_1 = \{C(t)x_n; x \in D, n \geq n(x), 0 \leq t < \epsilon_n(x)\}.$$

Since $D_0 \subset D_1$, D_1 is dense in X, and as before, if $y \in D_1$, there exists $\epsilon'(y) > 0$ such that $y \in D(C(t))$ for $|t| < \epsilon'(y)$. By (1),

$$2u^{-2}(C(u)y - y, y) \geq 0 \qquad (|u| < \epsilon'(y)). \qquad (5)$$

Writing $y = C(t)x_n$ for some $n \geq n(x)$ and some $t \in [0, \epsilon_n(x))$, we have

$$2u^{-2}[C(u)y - y] = u^{-2}[2C(u)C(t)x_n - 2C(t)x_n]$$

$$= u^{-2}[C(t+u) + C(t-u) - 2C(t)]x_n. \qquad (6)$$

By (4), the last expression has a (strong) limit as $u \to 0$, which we denote $A_0 y$. The operator A_0 is linear on the dense domain D_1, and positive (by (5) and (6)), i.e.,

$$(A_0 y, y) \geq 0 \qquad (y \in D_1).$$

Let A be the Friedrichs selfadjoint extension of A_0 (cf. Theorem XII.5.2 in [DS,II]), and let E be its resolution of the identity. Denote $E_m = E([0, m])$ and $A_m = E_m A = \int_0^m sE(ds)$ for $m \in \mathbb{N}$. Note that A_m is a *bounded* positive (selfadjoint) operator. Let $x_n \in D_0$; for $|t| < \epsilon_n(x)$,

$$\frac{d^2}{dt^2} E_m C(t)x_n = E_m A_0 x_n = A_m x_n$$

125

by (4) and the definition of A_0.

From the spectral representation, $\cosh(tA_m^{1/2})E_m x_n$ is also a solution of

$$v'' = A_m v, \quad v(0) = E_m x_n, \quad v'(0) = 0.$$

By the uniqueness of the solution, we have

$$E_m C(t)x_n = \cosh(tA_m^{1/2})E_m x_n = \cosh(tA^{1/2})E_m x_n$$

for $|t| < \epsilon_n(x)$.

When $m \to \infty$, $E_m C(t)x_n \to C(t)x_n$ (for each $|t| < \epsilon_n(x)$); in particular, $E_m x_n \to x_n$. Also

$$\cosh(tA^{1/2})E_m x_n = E_m C(t)x_n \to C(t)x_n$$

(when $m \to \infty$).

Since $\cosh(tA^{1/2})$ is closed, it follows that $x_n \in D(\cosh(tA^{1/2}))$ and

$$\cosh(tA^{1/2})x_n = C(t)x_n \tag{7}$$

for $x_n \in D_0$ and $|t| < \epsilon_n(x)$.

For $n \to \infty$, $x_n \to x$, and by (3),

$$C(t)x_n = (1/2)\int_{\mathbb{R}} h_n(s)[C(t+s) + C(t-s)]x\,ds \to C(t)x$$

for $|t| < \epsilon(x)$.

Since $cosh(tA^{1/2})$ is closed, it follows from (7) that x is in its domain, and

$$C(t)x = \cosh(tA^{1/2})x \tag{8}$$

for all $x \in D$ and $|t| < \epsilon(x)$. The uniqueness of A is proved as follows. If B is also a positive selfadjoint operator satisfying the identity in the theorem, and if E and F are the resolutions of the identity for A and B respectively, then

$$\cosh(tA^{1/2})x = \cosh(tB^{1/2})x \quad (x \in D, |t| < \epsilon(x)).$$

Since $|\cosh z| \le \cosh(\Re z)$, the above identity (written in term of the corresponding spectral integrals) extends analytically to t complex in the strip $|\Re t| < \epsilon(x)$. In particular for $t \in i\mathbb{R}$, we have $\cos(sA^{1/2})x = \cos(sB^{1/2})x$ for all $x \in D$, hence for all $x \in X$ by density (since the operators are *bounded*), and for all $s \in \mathbb{R}$. Thus

$$\int_0^\infty \cos(su^{1/2})E(du)x = \int_0^\infty \cos(su^{1/2})F(du)x$$

for all $x \in X$ and $s \in \mathbb{R}$. By the uniqueness property of the cosine transform, it follows that $E = F$, and therefore $A = B$.||||

We consider next the general case of a local cosine family of symmetric operators.

The following notation will be used. If A is a selfadjoint operator, and E is its resolution of the identity, we let

$$A^+ := \int_0^\infty u E(du); \quad A^- := - \int_{-\infty}^0 u E(du)$$

with the usual domains.

The cartesian product $X^2 := \{[x, y]; x, y \in X\}$ is considered as a Hilbert space with the inner product $([x, y], [x', y']) := (x, x') + (y, y')$.

If T is an operator on X with domain $D(T)$, we let

$$\mathbb{T}[x, y] := [Tx, -Ty] \qquad ([x, y] \in D(T)^2).$$

2.44. LEMMA. If T is symmetric, then $\mathbb{T}$ has a selfadjoint extension.

PROOF. Let $J[x, y] := [y, -x]$. Then J is unitary, $J^2 = I$ (the identity operator on X^2), and $JD(\mathbb{T}) = D(\mathbb{T})$. One verifies that

$$\mathbb{T} - iI = J(\mathbb{T} + iI)J.$$

Therefore

$$[ran\,(\mathbb{T} - iI)]^\perp = J[ran\,(\mathbb{T} + iI)]^\perp.$$

This implies that $\mathbb{T}$ (which is obviously symmetric) has equal deficiency indices $(n_- = n_+)$, and has therefore a selfadjoint extension (cf. [DS,II], Chapter XII).||||

2.45. THEOREM. Let D be a dense linear manifold in the (complex) Hilbert space X, and let $C(.)$ be a local cosine family of symmetric operators on D. Then there exists a selfadjoint operator A on X such that

$$C(t)x = \cosh[t(A^+)^{1/2}]x + \cos[t(A^-)^{1/2}]x$$

for all $x \in D$ and $|t| < \epsilon(x)$.

PROOF. As in the proof of Theorem 2.43, we may assume that each $C(t)$ is closed. With notation as in the proof of that theorem, we obtain the operator A_0 defined on D_1 (up to that point, the "bounded below" hypothesis was *not* used). Since

$$(C(u)y - y, y) = (y, C(u)y - y) \qquad (y \in D_1, |u| < \epsilon'(y)),$$

it follows from (4) and (6) that A_0, with the dense domain D_1, is *symmetric*. Let $\mathbb{A}$ be a selfadjoint extension of the operator $\mathbb{A}_0$ associated with A_0 (cf. Lemma 2.44), and let E be its resolution of the identity. Consider the projections $E_m^+ := E([0, \infty))$ and the *bounded* positive selfadjoint operators $A_m^+ := E_m^+ \mathbb{A}$ for $m \in \mathbb{N}$.

For $x_n \in D_0$, let

$$\xi_{nm}^+(t) := E_m^+[C(t)x_n, 0] \qquad (|t| < \epsilon_n(x)).$$

Since $[C(t)x_n, 0] \in D_1^2 = D(\mathbb{A}_0)$, $\mathbb{A}_0 \subset \mathbb{A}$, and the projection E_m^+ commutes with $\mathbb{A}$, it follows from (4) and the definition of A_0 that

$$\frac{d^2}{dt^2}\xi_{nm}^+(t) = E_m^+[A_0 C(t)x_n, 0] = E_m^+\mathbb{A}_0[C(t)x_n, 0]$$

$$= E_m^+\mathbb{A}[C(t)x_n, 0] = E_m^+\mathbb{A}E_m^+[C(t)x_n, 0] = A_m^+\xi_{nm}^+(t)$$

for $|t| < \epsilon_n(x)$. By uniqueness of the solution of the second order ACP, we then have for all $|t| < \epsilon_n(x)$:

$$\xi_{nm}^+(t) = \cosh[t(A_m^+)^{1/2}]\xi_{nm}^+(0)$$

$$= \cosh[...]E_m^+\xi_{nm}^+(0) = \cosh[t(A^+)^{1/2}]\xi_{nm}^+(0). \qquad (9)$$

When $m \to \infty$,

$$\xi_{nm}^+(t) \to E([0, \infty))[C(t)x_n, 0] := \xi_n^+(t)$$

for $|t| < \epsilon_n(x)$. In particular, $\xi_{nm}^+(0) \to \xi_n^+(0)$, and $\cosh[t(A^+)^{1/2}]\xi_{nm}^+(0) \to \xi_n^+(t)$ by (9). Since $\cosh[...]$ is closed (it is selfadjoint!), it follows that $\xi_n^+(0)$ belongs to its domain, and

$$\cosh[t(A^+)^{1/2}]\xi_n^+(0) = \xi_n^+(t) \qquad (|t| < \epsilon_n(x)). \qquad (10)$$

Similarly, letting $E_m^- = E([-m.0])$, etc..., we find that

$$\cosh[t(-A^-)^{1/2}]\xi_n^-(0) = \xi_n^-(t) \qquad (|t| < \epsilon_n(x)), \qquad (11)$$

where $\xi_n^-(t) := E((-\infty, 0])[C(t)x_n, 0]$.

Adding (10) and (11), we obtain

$$[C(t)x_n, 0] = \cosh[t(A^+)^{1/2}]\xi_n^+(0) + \cos[t(A^-)^{1/2}]\xi_n^-(0) \qquad (12)$$

for $|t| < \epsilon_n(x)$.

Let P denote the orthogonal projection of X^2 onto X (*identified* with its imbedding in X^2 as $\{[x, 0]; x \in X\}$). Then P commutes with $\mathbb{A}_0$, hence with

128

A, and so $A := P\mathbb{A}$ is a selfadjoint operator on X with resolution of the identity $PE(.) = E(.)P$. We have

$$\cosh[t(\mathbb{A}^+)^{1/2}]\xi_n^+(0) = \int_0^\infty \cosh(tu^{1/2})E(du)[x_n, 0]$$

$$= \int_0^\infty \cosh(tu^{1/2})[E(du)P][x_n, 0] = \cosh[t(A^+)^{1/2}]x_n,$$

and similarly for the second term in (12) (we have identified $[x_n, 0]$ with x_n). Thus

$$C(t)x_n = \cosh[t(A^+)^{1/2}]x_n + \cos[t(A^-)^{1/2}]x_n \tag{13}$$

for all $|t| < \epsilon_n(x)$.

Recall now that for each $x \in D$, $x_n \to x$ and $C(t)x_n \to C(t)x$ when $n \to \infty$ (for $|t| < \epsilon(x)$). Since the operator on the right of (13) is closed, it follows that x belongs to its domain and the identity in Theorem 2.45 is verified.||||

The following concept corresponds to that of analytic vectors in the context of cosine families.

2.46. DEFINITION. A **semianalytic vector** for A is a vector $x \in D^\infty(A)$ such that

$$\Sigma_{n=0}^\infty \frac{t^{2n}}{(2n)!}||A^n x|| < \infty$$

for some $t > 0$ (depending on x).

If A is positive, we may immitate the proof of Theorem 2.39, applying Theorem 2.43 to the local cosine family of *bounded below* operators

$$C(t)x = \Sigma_{n=0}^\infty \frac{t^{2n}}{(2n)!}A^n x,$$

to obtain the following result:

2.47. THEOREM (Nussbaum's Semianalytic Vectors Theorem). Let A be a positive operator with a *dense* set of semianalytic vectors. Then A is selfadjoint.

The details of the proof are omitted. Note that if A is not assumed to be closed, the conclusion is that A is essentially selfadjoint.

NOTES and REFERENCES

PART I. GENERAL THEORY.

The standard books on semigroups are [D, G, HP, P], with chapters in general texts like [DS I-III, Kat1, RS].

The Hille-Yosida space. The terminology and Theorem 1.23 are from [K5].

Semigroup convergence. Theorem 1.32 goes back to [Tr].

Exponential formulas. The treatment here follows [D,P], and is based on work by [Kat3, C1, C2, Tr].

Perturbations. Theorem 1.38 is due to Hille-Phillips. The proof given here is basically the one in [DS1].

Groups. Theorem 1.40 is from [N]. Theorem 1.41 is the classical Stone theorem. Theorem 1.49 and the following analysis are from [K3].

Analyticity. Theorem 1.54 is from [Liu], but the short proof given here is new.

Non-commutative Taylor formula. The results of this section are from [K7].

PART II. GENERALIZATIONS.

Pre-semigroups. The concept appears in germinal form in [DP] (under the name of regularizable semigroups). In [DPg], the name "C-semigroup" is coined, and the detailed analysis of these families is started (see [DL1-DL3, M1, M2, MT1-MT3, T1, T2], as a partial list for this subject). Since a C-semigroup is *not* a semigroup (unless $C = I$), we prefered to call it here a *pre-semigroup*.

Theorems 2.3-2.5 are from [DL1].

Theorem 2.8 is from [DL2] (but we coined the term "exponentially tamed" as a reference to Property 3.).

The Semi-simplicity manifold. The concept goes back to [K1] for a single *bounded* operator, with extensions to unbounded operators appearing in [K1, K2, KH2, KH3]. Theorem 2.11 is from [KH2]. A variant of this theorem is found in [KH3]. Theorem 2.20 is from [K2] (see also [K4]). Lemma 2.24 is from [KH3] (see also [DLK]). The concepts of the Laplace-Stieltjes space and of the Integrated Laplace space for a family of closed operators were introduced and studied in [DLK]. Theorems 2.23, 2.28, 2.29, and 2.31 are from [DLK] (with some modification of the proofs). Theorem 2.36 is a special case of the main result of [DL3]. Integrated semigroups were introduced in [Neu].

Semigroups of unbounded symmetric operators.

First results on this subject were obtained in [De] and [Nus]. A general theory of semigroups of unbounded operators in Banach space was developed in [H1, H2]. Theorem 2.37 is from [KL], as well as the proof of Theorem 2.39 (which appeared originally in [Nel], with a different proof). Another proof of Theorem 2.37 is found in [Fr], and serves a model for the proofs of Theorems 2.43 and 2.45 (first published in [KH3]). The concept of "semianalytic vector" is due to Nussbaum, as well as Theorem 2.47 (with a proof independent of the result on local cosine families; see [RS]). The results on local semigroups are generalized to a Banach space setting in [KH1] (see also [K4]). For cosine families of closed operators, a "semi-simplicity manifold" can be constructed as in [KH3] to provide a spectral integral representation, as Theorem 2.28 does it for semigroups of closed operators (cf. Theorem 4.2 in [KH3], with the obvious modifications needed in Definition 4.1 and in the proof of the theorem).

BIBLIOGRAPHY

[A1] Arendt, W., Resolvent positive operators, Proc. London Math. Soc. 54 (1987), 321-349.

[A2] Arendt, W., Vector valued Laplace transforms and Cauchy problems, Israel J. Math. 59 (1987), 327-352.

[Bo] Bochner, S., A theorem on Fourier-Stieltjes integrals, Bull. Amer. Math. Soc. 40 (1934), 271-276.

[BZ1] Burnap, C. and Zweifel, P.F., A note on the spectral theorem, Integral Equations and Oper. Theory 9 (1986), 305-324.

[BZ2] Burnap, C. and Zweifel, P.F., Cauchy problems involving non-selfadjoint operators, Appl. Anal. 25 (1987), 301-318.

[C1] Chernoff, P.R., Note on product formulas for operator semigroups, J. Funct. Anal. 2 (1968), 238-242.

[C2] Chernoff, P.R., Product formulas, nonlinear semigroups, and addition of unbounded operators, Memoir Amer. Math. Soc. 140' Providence, R.I., 1974.

[DP] Da Prato, G., Semigruppi regolarizzabili, Ricerche Mat. 15 (1966), 223-248. (1966), 2

[D] E.B. Davies, One-Parameter Semigroups, Academic Press, London, 1980.

[DPg] Davies, E.B. and Pang, M.M., The Cauchy problem and a generalization of the Hille-Yosida theorem, Proc. London Math. Soc. 55 (1987), 181-208.

[DL1] deLaubenfels, R., C-semigroups and the Cauchy problem, J. Funct. Anal. 111 (1993), 44-61.

[DL2] deLaubenfels, R., C-semigroups and strongly continuous semigroups, Israel J. Math. 81 (1993), 227-255.

[DL3] deLaubenfels, R., Integrated semigroups, C-semigroups, and

the Abstract Cauchy problem, Semigroup Forum, to appear.

[DL4] deLaubenfels, R., Existence Families, Functional Calculi, and Evolution Equations, Lecture Notes in Mathematics, Vol. 1570, Springer-Verlag, Berlin-Heidelberg-New York, 1994.

[DLK] deLaubenfels, R. and Kantorovitz, S., Laplace and Laplace-Stieltjes space, J. Funct. Anal. 116 (1993), 1-61.

[DLK1] deLaubenfels, R. and Kantorovitz, S., The semi-simplicity manifold on arbitrary Banach spaces, J. Funct. Analysis, to appear.

[De] Devinatz, A., A note on semi-groups of unbounded self-adjoint operators, Proc. Amer. Math. Soc. 5 (1954), 101-102.

[DS I-III] Dunford, N. and Schwartz, J.T., Linear Operators, Part I-III, Interscience, New York, 1958, 1963, 1971.

[F1] Faris, W.G., The product formula for semigroups defined by Friedrichs extensions, Pac. J. Math. 22 (1967), 47-70.

[F2] Faris, W.G., Self-adjoint Operators, Lecture Notes in Math. Vol. 433, Springer, Berlin, 1975.

[Fat] Fattorini, H.O., The Abstract Cauchy Problem, Addison Wesley, Reading, Mass., 1983.

[Fr] Frohlich, J., Unbounded symmetric semigroups on a separable Hilbert space are essentially selfadjoint, Adv. Appl. Math. 1 (1980), 237-256.

[G] Goldstein, J.A., Semigroups of Operators and Applications, Oxford, New York, 1985.

[HiK] Hieber, M. and Kellermann, H., Integrated semigroups, J. Funct. Anal. 84 (1989), 160-180.

[HP] Hille, E. and Phillips, R.S., Functional Analysis and Semigroups, Amer. Math. Soc. Colloquium Publ. 31, Providence, R.I., 1957.

[H1] Hughes, R.J., Semigroups of unbounded linear operators in Banach space, Trans. Amer. Math. Soc. 230 (1977), 113-145.

[H2] Hughes, R.J., On the convergence of unbounded sequences of semigroups, J. London Math. Soc. (2), 16 (1977), 517-528.

[H3] Hughes, R.J., A version of the Trotter product formula for quadratic-form perturbations, J. London Math. Soc. (2), 30 (1984), 322-334.

[HK] Hughes, R.J. and Kantorovitz, S., Boundary values of holomorphic semigroups of unbounded operators and similarity of certain perturbations, J. Funct. Anal. 29 (1978), 253-273.

[K1] Kantorovitz, S., The semi-simplicity manifold of arbitrary operators, Trans. Amer. Math. Soc. 123 (1966), 241-252.

[K2] Kantorovitz, S., Characterization of unbounded spectral operators with spectrum in a half-line, Comment. Math. Helvetici 56 (1981), 163-178.

[K3] Kantorovitz, S., Spectrality criteria for unbounded operators with real spectrum, Math. Ann. 256 (1981), 19-28.

[K4] Kantorovitz, S., Spectral Theory of Banach Space Operators, Lecture Notes in Math., Vol. 1012, Springer, Berlin-Heidelberg- New York, 1983.

[K5] Kantorovitz, S., The Hille-Yosida space of an arbitrary operator, J. Math. Anal. Appl. 138 (1988), 107-111.

[K6] Kantorovitz, S., Sur le calcul fonctionnel dans les algebres de Banach, C.R. Acad. Sci. Paris, 317 (1993), 951-953.

[K7] Kantorovitz, S., C^n-operational calculus, non-commutative Taylor formula, and perturbation of semigroups, J. Funct. Anal. 113 (1993), 139-152.

[K8] Kantorovitz, S., On Liu's analyticity criterion for semigroups, Semigroup Forum, to appear.

[KH1] Kantorovitz, S. and Hughes, R.J., Spectral representation of local semigroups, Math. Ann. 259 (1982), 455-470.

[KH2] Kantorovitz, S. and Hughes, R.J., Spectral representation for unbounded operators with real spectrum, Math. Ann. 282 (1988), 535-544.

[KH3] Kantorovitz, S. and Hughes, R.J., Spectral analysis of certain operator functions, J. Operator Theory 22 (1989), 243-262.

[Kat1] Kato, T., Perturbation Theory for Linear Operators, Springer- Verlag, New York, 1966.

[Kat2] Kato, T., A characterization of holomorphic semigroups, Proc. Amer. Math. Soc. 25 (1970), 495-498.

[Kat3] Kato, T., Trotter's product formula for an arbitrary pair of self-adjoint contraction semigroups, Topics in Functional Analysis, Adv. in Math. Supplementary Studies I, Acad. Press, New York, 1978 (185-195).

[KL] Klein, A. and Landau, L.J., Construction of a unique selfadjoint generator for a symmetric local semigroup, J. Funct. Anal. 44 (1981), 121-137.

[L] Lions, J.L., Semi-groupes distributions, Portugalae Math. 19 (1960), 141-164.

[Liu] Liu, Y., An equivalent condition for analytic C_o-semigroups, J. Math. Anal. Appl. 180 (1993), 71-78.

[M1] Miyadera, I., On the generators of exponentially bounded C-semigroups, Proc. Japan Acad. 62 (1986), 239-242.

[M2] Miyadera, I., A generalization of the Hille-Yosida theorem, Proc. Japan Acad. 64 (1988), 223-226.

[MT1] Miyadera,I. and Tanaka, N., Exponentially bounded C-semigroups and integrated semigroups, Tokyo J. Math. 12 (1989), 99-115.

[MT2] Miyadera, I. and Tanaka, N., Exponentially bounded C-semigroups and generation of semigroups, J. Math. Anal. Appl. 143 (1989). 358-378.

[MT3] Miyadera, I. and Tanaka, N., A remark on exponentially bounded C-semigroups, Proc. Japan Acad. Ser. A 66 (1990), 31-35.

[N] Nagy, B.Sz., On uniformly bounded linear transformations in Hilbert space, Acta Sci. Math. (Szeged) 11 (1947), 152-157.

[Nel] Nelson, E., Analytic vectors, Ann. Math. 70 (1959), 572-615.

[Neu] Neubrander, F., Integrated semigroups and their application to the abstract Cauchy problem, Pacific J. Math. 135 (1988), 111-155.

[Nus] Nussbaum, A.E., Spectral representation of certain one-parameter families of symmetric operators in Hilbert space, Trans. Amer. Math. Soc. 152 (1970), 419-429.

[P] Pazy, A., Semigroups of Linear Operators and Applications to Partial Differential Equations, Springer, New York, 1983.

[RS] Reed, M. and Simon, B., Methods of Modern Mathematical Physics II, Acad. Press, New York, 1975.

[R1] Rudin, W., Real and Complex Analysis, McGraw-Hill, New York, 1966.

[R2] Rudin, W., Fourier Analysis on Groups, Interscience Publishers, New York, 1962.

[SW] Stein, E.M. and Weiss, G., Introduction to Fourier Analysis on Euclidean Space, Princeton University Press, Princeton, 1971.

[T1] Tanaka, N., On the exponentially bounded C-semigroups, Tokyo J. Math. 10 (1987), 107-117.

[T2] Tanaka, N., Holomorphic C-semigroups and holomorphic semigroups, Semigroup Forum 38 (1989), 253-263.

[Tr] Trotter, H.F., Approximation of semi-groups of operators, Pacific J. Math. 8 (1958), 887-919.

[W1] Widder, D.V., Necessary and sufficient conditions for the representation of a function by a doubly infinite Laplace integral, Bull. Amer. Math. Soc. 40 (1934), 321-326.

[W] Widder, D.V., The Laplace Transform, Princeton University Press, Princeton, 1941.